ARTISTS-IN-LABS

NETWORKING IN THE MARGINS

Editor: Jill Scott

artistsinlabs.ch

z ___ hdk

Zürcher Hochschule der Künste

SpringerWienNewYork

Editor
Jill Scott

Zurich University of the Arts (ZHdK), Switzerland
Institute for Cultural Studies in the Arts

Produced by artists-in-labs
Co-directors
Irène Hediger and Jill Scott

Research Assistant
Alain Rickli

Graphic Design
Karin Schiesser

Cover Illustration
Karin Schiesser

Copy Editors
Susanne N. Hillman, Juanita Schläpfer-Miller

Documentary Films
Marille Hahne

Images on cover
Christian Gonzenbach, Sylvia Hostettler, Alina Mnatsakanian,
Ping Qiu, Hina Strüver & Mätti Wüthrich, Claudia Tolusso

© 2010 Springer-Verlag/Wien
Printed in Austria

SpringerWienNewYork is part of
Springer Science+Business Media
springer.at

Printed on acid-free and chlorine-free bleached paper
SPIN: 80015025

With 107 colour figures and 73 black and white figures

Library of Congress Control Number: 2010929865

ISBN 978-3-7091-0320-3 SpringerWienNewYork

ESSAYS

CASE STUDIES

LIFE SCIENCES

COGNITION & PHYSICS

COMPUTING & ENGINEERING

BIOGRAPHIES
DVD AND ANALYSES

FOREWORD
RESEARCH: SEARCH AGAIN Sigrid Schade

Since the early sixties, when alternative creative centres were founded in reaction to C.P. Snows theory of the Two Cultures, the interaction between artists and scientists has been staged on some old assumptions. These historical attributes constituted the artists 'subjectivity': the creative production of things not absolutely needed but something new, spiritual, intuitive, unconventional, critical, marginalized and/or heroic. The historically ascribed qualities of the scientist were to be objective, follow systematic methodologies and deliver the practical tools and products that a (post) capitalist and technologically dependent society wanted or questioned them about. These staged encounters seemed to signal themselves in that they offered an 'agora' for a complementary 'clash of civilizations'. The hope was that either both sides would profit or that both disciplinary creativities would be improved so that their productions would be even more relevant to society. But for whom were they relevant exactly? The answers given were absolutely dependent on the cultural and political framework within which the experimental setting was placed.

Today, these complementary constructed discourses on art and sciences have collapsed. True, there might still be scientific discourses, which only rely on the grounds of empirical truth, and the art market still celebrates the concept of authorship, however, both sides have changed. Within history and social studies of science it is acknowlegded, that the knowledge which empirical research provides is dependent on the specific methods and procedures it uses. Similarly, as in other discourses of knowledge in the humanities and cultural studies, the cultural, political, practical and theoretical setting of the research and its framing greatly affects the results. As Karin Knorr Cetina suggests, scientists don't search or find knowledge or truth, they actually make it (Knorr Cetina 1999). Currently, art universities are struggling with their 'own' concepts of what constitutes 'artistic' research or 'designerly ways of knowing'. These discourses are sometimes forced by funding institutions that traditionally delt with scientific research. Trying to close a legitimation gap in society does not always show the level of self-reflection, which still should be a basic principle of art, design and media research today. Any discourse about art concepts or practices may need this reflection in order to be called artistic research (Mareis 2010). In Switzerland, there is still a problematic differentiation between basic and applied research and the art schools have often been forced to deal with the self-declaration of their own artistic methodologies. This not only produces new problems but it does not acknowledge the current changes in the sciences themselves. Recently, Mieke Bal examined these kind of problems and effects in 'Travelling Concepts' (Bal 2002) and suggested that they could produce errors in translations, distortions and misunderstandings, particularly as concepts often change their meaning as soon as they are transferred from one epistemic field into another.

The Institute for Cultural Studies in the Arts at the Zurich University of the Arts has supported the Swiss artists-in-labs-program since 2006. The major sponsor, the Bundesamt für Kultur (Swiss Federal Office of Culture) is absolutely aware of these developments in artistic and scientific research. The Program offers a reflective space in which the encounters of artists and scientists share fantasies or illusions of knowledge and responsibilities concerning social and political change. The co-directors Prof. Dr. Jill Scott and Irène Hediger provide a context in which the cultural settings of these experiments become visible. By juxtaposing people, concepts and procedures, the Swiss artists-in-labs program enables both sides to mutually identify the interesting, the fascinating but also the problematic sides of each other.

References

Knorr, C 1999, E*pistemic Cultures: How the Sciences Make Knowledge,* Cambridge Massachusetts.

Mareis, C 2010, *Design als Wissenskultur. Interferenzen zwischen Design- und Wissensdiskursen seit 1960,* transcript, Bielefeld, forthcoming.

Bal, M 2002, *Travelling Concepts in the Humanities. A Rough Guide,* University of Toronto Press, Toronto.

INTRODUCTION AND RELATED ESSAYS

INTRODUCTION:
NETWORKING IS BOTH AN ART AND A SCIENCE! Jill Scott

Artists-in-labs: Networking in the Margins is the second volume in this series of cultural studies, that blurs the boundaries between art and science research. This book presents case studies from 2007 to 2009 by 12 artists from the disciplines of sculpture, installation, performance and sound, in collaboration with scientists from the fields of physics, computer technologies, environmental ecology, neuroscience and psychology. While artists have become more involved in ethical and social debates about scientific discovery, scientists have been exposed to the processes and contexts of art. Thus, networking in the margins of art and science tends to expand the borders of the exact sciences and to contribute to a more robust level of dialogue within the humanities and the arts. These trans-disciplinary potentials are contextualized by 9 prominent authors, who each shed light on the future implications of such interactions for society.

In any traditional essay, margins are often used for reflective additions or to make notes, which criticize the main text or offer surprising comments or insights by a reader, notes which are associative and relevant for further research. This book constitutes such a margin, because it is full of note-taking by scientists, artists, theorists and even filmmakers and this margin is growing to such an extent that finally researchers can make comparisons. Marginalisation is also the social process of becoming, or being made marginal and marginalized individuals can also be excluded from services, programs, and policies (Young 2000). Certainly, the contemporary arts have a history of being marginalized by the science community and vice versa. This marginalisation has come about because of three levels: a lack of respect, a specialized focus embedded in disciplinary institutions, and the differences in commercial structure and funding. In both art and science contexts, individuals have often been excluded from each others services, programs, and policies. However, from a sociological point of view, marginalisation is also the social process of being excluded and material deprivation is the most common result. Consequently, both disciplines are also marginalized by society, a fact utilized by the cultural mainstream.

Instead, the writers in this book recognize the potential to create a complimentary intersection in between these margins, and investigate their capacity to overlap with each other. Artists and scientists have already been networking because they want to start 'thinking outside of the box'! Residency programs are encouraging artists who are interested in research to witness the process of scientific discovery and science labs are becoming interested in the art process. These science hosts are not dominant mainstream labs with public acceptance and millions of funds behind them, but university learning centres where knowledge can easily be shared with 'outsiders'. Thus the representatives from each field already have alternative approaches, coupled with interests to explore new methods of communication and to observe each production and discovery process. Usually they share a healthy ethical attitude about how to communicate with the public and a need to form trans-disciplinary teams for new ideas. By providing a context where the close observation of art-making can unfold, perhaps unprecedented liaisons and novel reflections on creativity and innovation can also develop.

The essays and documents in this book are focused on an exchange of knowledge between the cultural sphere of the arts and the cultural sphere of science, through the sharing of discourses, aesthetics, tools, methodologies, and exhibitions. Each writer has generated an innovative approach, encouraging scientifically robust artworks and the opening up of critical debates for the general public. Each author comments on the creative and communicative potentials of the disciplines involved and on how experience can shift the roles of artists and scientists in the future.

In her essay entitled *Formative encounters: Laboratory life and artistic practice* the sociologist Andrea Glauser gives an overview of the value of 'foreignness' between artists and scientists in sociological terms and shows how this aspect can be advantageous and cause shifts in new know-how transfer. This 'foreignness' can also create educational potentials that affect authorship and productivity, particularly in relation to the technological tools of science. She suggests that such an 'irritation' of stereotypes and patterns of tradition that a resident artist will encounter, because they are 'the outsider', will only be shifted by a mutual understanding of production processes and 'an appreciation of diverging perspectives, emotional reactions, varieties in inspiration and interpretation' from both sides, before future discourses can take place in the public realm.

In *On the nature of interactions*, botanist Lloyd Anderson also points to problems that stem from our tendency to construct dichotomies and stereotypes, and he compares Snow's *The Two Cultures'* theories with a third theory he defines as the Agora: a meeting place where both the highly specialized and the generalists can co-exist. This agora is also a place where 'one specialist can modify the environment that the other may operate in' just like in an ecological corridor. In these corridors, the effects of species and their actions inside this environment can not only cause interaction but also mediation. For Anderson, this is a margin where the illustration of science and the interpretation of science seems to breed a natural place of interdisciplinary sharing and reshaping of ideas, but one in which respect for our respective disciplines is an important factor to maintain. Here the main interactive challenge is to share in a 're-thinking and re-imagining of what is around us'.

In *Bacteria, robots and networks* the cultural curator Dominik Landwehr compares the fascination, that drives an artist to build robots, with that of a researcher in Artificial Intelligence. While an artist wants to communicate his fascination with machines to an audience, an AI scientist hopes to gain an understanding of intelligence by building hardware and software and provoking new questions for further research. Landwehr suggests that new ideas may not only come from mixing both fields together, but from acknowledging that analogue technology is still very bewitching for an audience. He cites The Enigma Machine and Turing's involvement in breaking the machine's encryption as paramount examples, and encourages a new enthusiasm about analogue technology. He uses examples of provisional robotic platforms by artists and their need to use workshops in order to explore the evolution of their concepts and to facilitate more intense dialogue. He applauds these methods because they promote intuition and association.

In *Artists-who-care!*, I took on the challenge of ethical and social content, specifically the sharing of this responsibility by artists and scientists. The essay is based on three claims: First, the number of contemporary artists who are more deeply committed and actively concerned about their inclusion in society is growing. Second, these same artists are often interested in the social impacts of scientific discovery and the ethical discourses surrounding the scientific process. Finally artworks embedded with this more robust interest in science and content may be able to help the general public shift their inter-relational equation between the sciences, the arts and the environment in which they live. These claims not only require the formation of organizations, which foster trans-disciplinary projects like the one featured in this book, but a general re-structuring of society's attitudes toward the role of culture.

By contrast, the next essay in this book, entitled *Art at the End of tunnel vision* is almost an artwork in itself. By utilizing the genre of creative prose, cultural philosopher Roy Ascott takes us on a journey of word-play that explores coherence between the discourses found in particle physics, technology, and consciousness. He actually suggests we adopt a 'technoetic' manifesto called 'syncretism': a state of mind, which will surely lead us out of the 'postmodern blind alley'. Throughout the text the plural pronouns of 'we' and 'our' are used to situate the reader into the text as an accomplice or new member of an evolved concept. Because, as David Bohm suggests, thought is a participatory activity. Ascott's manifesto calls for an evolution of participatory thought, one which includes syncretism and coherence between

disciplines and an understanding of variable realities where the boundaries are permeable. 'Syncretism', he suggests, can help to achieve a 'reconstruction and regeneration of the world' as we know it and offer us the responsibility to 're-design ourselves'. The final essay talks about the challenges of sharing methodologies between disciplines. In *Art and science research teams? Some arguments in favour of a culture of dissent*, art historian Nina Zschocke outlines the variables within those instructions, models, and methodologies given to the test participants, rather like those given to actual empirical results. She actually supports a clash of interests because this could offer a critique of the host's assumptions and practices, and also increase the chance that the interaction might produce more unpredictable artistic results. Artists can contribute to their own field of artistic research because they create new meaning, 'make things strange', culturally 'hack', play with the act of observation, network by strolling freely throughout the spaces of knowledge production, hunt and gather, or ask unexpected questions like 'who is concerned' and 'what is considered to be an issue?' Artists should not enter a science lab to use the lab tools or to 'be a parasite' but to see scientific machines as metaphors for all kinds of systems, explore scientific theories as compositional strategies, and not ignore the disturbances and imbalances, mutations and reproductions, and errors that may be possible. It would indeed be sad if they became scientists, directly transferring methods, copying their orders and innovation processes.

The case studies and films form the next part of this book. Beginning with an introduction by Irène

Hediger (co-director of the artists-in-labs program) a series of in-depth reflections are raised by artists and scientists: What can an artist learn from being immersed inside a science lab? How could this experience be applied to the artist's own matters of concern? How can the resultant artwork relate to the original scientific research in the labs themselves, as well as invite the audiences to be involved? What do scientists learn from meeting artists and watching their processes unfold? Here, through artist diaries, short introductions, lab reports, and filmed documentary evidence, the reader can not only compare and synthesise the strengths of these investigations which have taken place over the last three years, but the curator or organizer can investigate the foundations needed to start their own artists-in-labs programs elsewhere. For embodied within the blurred boundaries of the margins of the art scene, as well as those of scientific practice, lies a responsible attitude to further this transdisciplinary discourse. One which values citizen science, fantasy, fact, unexpected results and immersive education. The artists-in-labs program was generously supported by The Swiss Federal Office of Culture and the Institute for Cultural Studies at the Zurich University of the Arts (ZHdK).

We would also like to add our thanks to the institutions who contributed to the making of this book including the Migros Culture Percentage, Switzerland; The Film Department of the Zurich University of the Arts (ZHdK) and all the other individuals who have worked patiently to make it all worthwhile.

FORMATIVE ENCOUNTERS: LABORATORY LIFE AND ARTISTIC PRACTICE Andrea Glauser

In recent years, the relationship between the arts and sciences has provoked a remarkable stir as it has been explored in books, lectures, exhibitions and conferences. Fundamental discussions have transpired, examining 'art as science' or 'science as art', and particular constellations, such as Laurie Anderson's residency at NASA, have been prominent subjects in the media.[1] The relationship between the arts and sciences is fascinating because, although these fields differ, they share several common features, in particular an orientation towards innovation. Artists and scientists are often seen as society's creative core. From that viewpoint, considerations of artistic and scientific practices are typically combined in the hope of shifting boundaries of specific knowledge and of producing a decisive 'enlargement of the universe of human discourse' (Geertz 1993, p. 14). Significantly, at the opening of the Institut de Recherche et Coordination Acoustique/Musique IRCAM in Paris in 1977, the composer and program director, Pierre Boulez, characterized the cooperation of artists and engineers as a 'utopian marriage of fire and water' (Boulez 1986, p. 490).

The interest in rapport between art and science is also manifest in cultural policies. In the last few years, remarkable efforts have been made to bring artists, scientists, and engineers together and to foster collaborative work. These efforts have taken a variety of forms. Besides the foundation of relatively stable institutes and academic study programs, collaboration is primarily advanced through residencies, meaning specific temporary placements of artists in laboratories and, far less frequently, scientists in artistic environments. Artist-in-residence programs,

as instruments of cultural promotion, are generally justified by the argument that they provide artists with infrastructure, networking possibilities, and the chance of broadening their horizons through their personal presence in a foreign cultural context (Behnke et al. 2008; Glauser 2009). In comparison to programs that supply artists with rather classical studios abroad, whether situated in New York, Beijing, Paris, or Bangalore, residencies in scientific contexts are usually more specifically focused on encounters and collaboration between actors with different backgrounds. In view of this emphasis, programs that foster collaboration between artists and scientists are not only of vital interest to contemporary forms of cultural policy. They are associated in a broader sense with questions of new methods of knowledge production, knowledge and innovation policies and their implications for social differentiation.

Combining heterogeneous knowledge in a post-industrial society

A widely supported assertion in sociology states that a decisive feature of contemporary society is its differentiation into a variety of cultural fields with each field, such as science or art, characterized by a particular logic of its own while not being completely independent from the others. With the formation of different microcosms, experiences of foreignness may multiply within societies (Amann & Hirschauer 1997, p. 12). Functional differentiation involves processes of specification and globalization as well as increasing complexity. This is sometimes perceived as an omen of a world falling apart. Yet differentiation does not solely imply increasing dissimilarity but also involves an enhancement of possible interdependences and cross-references between cultural fields. Some statements even suggest that the interrelations between fields become increasingly important, which places the assumed primacy of functional differentiation into question. With regard to the relationship between art and science, Werner Rammert's assumptions related to a new regime of knowledge production and a 'fragmental type of differentiation' are especially interesting (Rammert 2003). Rammert argues that, with the end of the industrial society, shifts in the way in which knowledge is produced can be observed. The regime of a complementary and disciplinary-based specialized pattern of production, which is typical for a functionally differentiated society, is increasingly losing its predominant role. Consequently, forms of production that are based on a combination of heterogeneous elements, actors, and forms of knowledge gain importance. An important factor concerning the interconnection of different domains is identified in the emergence of new media and technologies (above all in the diffusion of the computer into almost all domains of life), which favor the formation of new research fields. Mixtures of epistemic cultures are widespread; transdisciplinary networks have emerged, and elements of all sorts of knowledge have been newly combined. Typically, the single elements in the combinations are functionally specialized forms of knowledge. Functionally specialized institutions and scientific disciplines remain as important factors, but they lose their privilege of exclusive access to the 'main stage' (Rammert 2003, p. 488). New domains emerge on the margins of specific fields. They react to disciplinary-related practices and institutional forms as well as to modes of cooperation. In the domain of engineering, such dynamics are nothing new, but this heterogeneous and hybrid model of knowledge production has now also reached the boundaries of traditional disciplines and fields. From a sociological point of view, it is, therefore, vital to concentrate attention both on the peculiarities of singular fields as well as on the interrelations and border crossings between fields.

In this essay, art-science cooperation is seen in the context of this relatively new form of knowledge production, which is on the one hand based on functionally differentiated structures – in Pierre Bourdieu's words, in the particular social games of art and science, the rules, codes and the specific form of *illusio* that each field produces (Bourdieu 1996). On the other hand it involves transformative dynamics through the combination of heterogeneous actors and forms of knowledge. The prime focus lies on the arts. First, the essay discusses why it is significant for artists to have access to scientific laboratories. As will be shown, such residencies are relevant to classical artistic demands and touch the core business of artistic production in a twofold way: they are vital to the use of new technologies for artistic purposes, especially to the development of new tools, and to artistic explorations of practices that are of central importance for current and future socio-cultural life. These two dimensions of artistic work are discussed historically and with reference to concrete constellations that have taken place in the context of the Swiss artists-in-labs program in recent years. Second, programs that bring artists and scientists together and foster collaborative work are discussed as structures that may transform the conception and perception of artistic work. Residencies in labs are sites where the artistic subject and the relations between artistic and scientific perspectives are debated and redefined. The studio-laboratory is a place of new forms of technological collaboration and artistic production that may extend practices either from the inside out or in the form of new views of them or demands on them. Different forms exist simultaneously; art & science collaborations should not be seen as a fusion, a general trend to dissolve boundaries. These programs do not reduce heterogeneity but, in combination with other (more or less institutionalized) forms of working alliances, further an interlocking of scientific and artistic practices.

The discussion of these questions is based on ethnographic research that includes interviews with artists and scientists, visits to laboratories and reviews of existing text and film documentaries and internet sources. The investigation rests upon the assumption that, in exploring the relationship between art and science, it is (at least temporarily) necessary to bracket the collective subjects 'art' and 'science' and to take specific constellations into account instead. This is because a) artistic and scientific fields are each in themselves highly differentiated; b) encounters between artists and scientists involve dynamics that are comparatively unique, that is, they are not (yet) consolidated in institutionalized forms; and c) encounters take place in highly different contexts.

Artists' residencies in labs as a central professional issue
In the artistic fields, such as music, the visual arts, etc., there are primarily two issues that make the possibility of rapport between art and science and residencies in laboratories a central professional concern. The first is the interest in inventing new tools and methods, that is, the appropriation of technologies, developed and used in other contexts, for artistic purposes. In this, the laboratory is primarily conceived as a site of interesting infrastructure and knowledge; access to scientific knowledge and respective technological infrastructure may help to extend the range of artistic articulation. In the context of the Swiss artists-in-labs stipends of the last years, this concern can be found, for example, in the residencies of Pablo Ventura at the Artificial Intelligence Laboratory at the University of Zurich, and Chandrasekhar Ramakrishnan's residency at the Computer Science Lab of the Swiss Federal Institute of Technology, also in Zurich. For the choreographer Pablo Ventura, interested in the interactions between man and machine in the context of dance, the residency in this lab included the opportunity of working together with experts in robotics and of exploring together with scientists the possibilities of dancing robots. Chandrasekhar Ramakrishnan, who works in the domain of composition and music software, developed a new programming language for performative multimedia artworks during his residency. In an interview about his residency, he points primarily to the possibility of discussing specific questions with scientists that have a vast experience in programming languages. The second issue of professional concern is that residencies in laboratories are potentially a vital professional source for artists because they allow the exploration of processes that are, in general, not easily accessible. Because life sciences, physics, computing, and engineering are of central

importance regarding current and coming forms of life, they are a highly interesting phenomenon for artistic reflection. Artists often work with such practices, using either a formal-aesthetic approach or a quasi-ethnographic mode (Foster 1996). From such perspectives, the laboratory is primarily conceived as a subject of investigation and the scientists, objects, and tools are seen as elements of a foreign universe that should be explored through artistic means. Comparable to the sociologists that have investigated the 'manufacture of knowledge' (Knorr Cetina 1981), 'epistemic cultures' (Knorr Cetina 1999), and 'laboratory life' (Latour & Woolgar 1979), artists have also become interested in laboratories and their material cultures. Significantly, Sylvia Hostettler, who was a visiting artist at the Centre for Integrative Genomics at the University of Lausanne in 2008, characterized her residency as a research journey. The Centre's investigation of the growth of model plants in specific light conditions in the domain of molecular biology became the starting point for her work on the 'apparent invisibility' of light reactions. Sylvia Hostettler inserted Petri dishes, a kind of apparatus used in scientific experiments, in her installations and so provoked reflections upon the cultural, and especially the aesthetic, dimensions of the lab's practices.[2] For Christian Gonzenbach, another visual artist and awardee at the Département de physique nucléaire et corpusculaire, University of Geneva in 2009, art is basically a form of investigation. In an interview, he explains that, although he is not very fond of ready-made strategies in the arts, it is almost impossible not to see the lab's infrastructure as a kind of ready-made installation. He identifies a central aspect of his presence at the Department of

Physics as the opportunity to gradually discover how physicists think and work. The artist's preoccupation with the scientific environment sometimes takes the form of concrete interventions within it – a strategy that has to be seen in the tradition of site-specific practices in the arts, common since the 1960s (Kwon 2004). In the context of the artists-in-labs program, an intervention in the material and substantial context of the hosting lab was undertaken by Ping Qiu, a visiting artist in 2008 at the Eawag Aquatic Research Centre in Dübendorf. She made interventions (among other projects) in the form of installations in the organization's area, working with toilets, which play a crucial role not only in Eawag's research activities but also in the history of art, due to Marcel Duchamp's legendary work. Her *Watercircle Toilet Fountain* and video installation *Toilet Mirror* associate references to these different contexts; they are at once a concrete, poetic and ironic comment on the affinities between art and science and the significance of water for life.[3]

Both the interest in developing new tools and the exploration of scientific practices imply keeping up with cultural and technological developments, by observing them and participating in their dynamics. This interest is not completely new; it is based on a historical yet still relevant conception of artistic work. The modern understanding of the artistic subject as bohemian, which was formulated in the 19th century in opposition to the predominant academic conception of artistic work on the one side and the capitalist entrepreneur on the other, involved criticism of specialization and the division of labor in modern society (Benjamin 1991, p. 556; Graña 1964). This point of view incorporates the conviction that artistic work is more than a specialized endeavor and that artists have to cross the boundaries of the artistic field to fulfill their 'missions' as artists. This paradoxical formulation can be found in almost pure form in the writings of French poets from the 19th century, especially in Baudelaire's text entitled *The Painter of Modern Life* of 1863. Here, the ideal artist is explicitly described as a cosmopolitan who is familiar with the manifold cultural forms and compelled by an incurable curiosity. The 'true' artist is characterized as an 'homme du monde [...] qui comprend le monde et les raisons mystérieuses et légitimes de tous ses usages' (Baudelaire 1976, p. 689). Of course today's concepts of the artistic profession do not completely correspond with these historical forms, yet the desire to be more than a specialist is nonetheless present. Especially in conceptual art strategies, which have been widespread since the 1960s, the boundaries of art are questioned, and the reflexive capacities of artistic practices are stressed. Significantly, the artists who have been participating in the artists-in-labs program in the last few years have often brought their practices into explicit relation with the traditions of conceptual art. It was an important context for the artists' interest in science and technology, in particular from a historical perspective (Barry, Born & Weszkalnys 2008, p. 38).

Transformative dynamics of art & science programs

Networking in the margins and residencies in scientific labs are more directly linked to basic artistic interests than is obvious at first sight. Although art-science collaborations are relevant to classical artistic demands, this fact does not imply conservative dynamics. On the contrary, the encounters have transformative potential in several respects. Residencies in labs are sites where artistic and scientific practices may be redefined. The encounters are likely to shape the knowledge, which is relevant for artistic practices, to effect a broadening of horizons and extend the 'space of possibles' in the artistic field (Bourdieu 1996, p. 234). From a sociological point of view, residencies in labs seem to be a highly promising starting point for crossing the borders of specific knowledge: Their personal presence in the lab allows the artists to observe scientific culture as 'lived order' (Pollner & Emerson 2001, p. 119). Furthermore, not only do residencies in labs offer the chance for artists to learn about scientific environments but also for scientists to gain insight into artistic problems, interests and strategies. Certainly, nine months of co-presence contains some limitations, as in the case of the artists-in-labs program. This time cannot substitute for a complete education or for the reading of publications in the fields in question. However, the importance of the actors' proximity upon which this program is based should not be underestimated. Even in times of globalized communication and the undoubted relevance of communication technologies that easily span spatial distances, there is no general loss of significance for proximity between actors. As several studies on the spatial dimensions of social relations point out, proximity of actors is almost indispensable to certain processes because it allows for medial complexity (Greve & Heintz 2005, p. 12; Stichweh 2000, p. 117). The need for medial complexity is especially relevant to processes that cannot rely on routine communication, for example the initiation of research projects. When research projects are already running, the researchers may work while scattered around the globe; however, it is almost impossible to begin a project without interactive co-presence. The same is undoubtedly true for delicate political negotiations and in diplomacy. Typically, the proximity of the involved actors and the enabled medial complexity is understood as a necessary pre-condition. This aspect has to be taken into account in the artists-in-labs' constellation as well. Irrespective of whether artists are primarily interested in developing new tools or in exploring scientific practices, their presence in the lab and interactive co-presence is a vital precondition to exchanges on specific, problem-oriented questions. This again is important to furthering the understanding of problems that are based on implicit, scarcely conscious knowledge but which can be made explicit through personal communication and directed inquiry. Personal communication plays a crucial role in learning processes (Heintz 1993, p. 223). Due to this constellation, residencies in labs allow for a kind of fieldwork that is reflected in the work. Furthermore, the work done in the context of the lab typically involves the apparatus and knowledge of scientists and technicians. Artistic practice is formed by the use of the instruments that are part of the laboratory. Both the invention of new tools and the mapping of cultural practices demand a great deal in the appropriation of relevant knowledge. The work developed during the residency involves highly unique mixtures of knowledge. It not only challenges the practices of restoration in the domain of new media, a domain that is also characterized by a combination of artistic and scientific knowledge, which has to keep in step with new forms of artistic production. But, in a fundamental way, it challenges the scope and borders of artistic positions with regards to questions of methods and subjects of investigation. As Howard S. Becker (1974, p. 770) states, art is a form of collective action: 'Relations of cooperation and constraint [...] penetrate the entire process of artistic creation and composition.' New forms of collaboration are therefore likely to reshape 'the space of possibles' in the artistic field, given the assumption that single artworks can only be understood in the context of other practices, in relation to other artistic 'position-takings' (Bourdieu 1996, p. 231).

However, artistic practice and identity are not only transformed by the way that artists use knowledge, instruments, and sites, which are made accessible by residency programs; the conception of art is in a certain way also defined by the scientists' perception of and reference to artistic practices. Through endeavors to bring artists and scientists together, these actors become mutually involved. The perception of the artists-in-labs and their practices are highly varied and heterogeneous. The differences are at least partly linked to differentiations within the scientific field and to the character of practical and theoretical problems in its subfields and particular disciplines. In some domains, such as computer science and artificial intelligence, collaborations between artists and scientists already have a basic tradition stemming from the 1960s. Before the breakthrough and broad diffusion of the personal computer, the collaboration sites were primarily universities, research centers, and radio stations that possessed the necessary infrastructure (Born 1995). In recent years, the collaborations became more decentralized. In the context of the artists-in-labs program, scientists who support artists in the development of new tools often perceive art as an area of application. Indeed, now and then artwork appears in scientific papers in the context of proving concepts. Sometimes, projects are worked out together and the artistic and technical aspects become tightly interwoven which can hardly be distinguished. Thus, multiple author-ships seem appropriate. In certain constellations, the artist's knowledge can be quite directly relevant to the scientific work. For scientists working in the domain of robotics, a dancer and choreographer's knowledge of the human body and bodily movement may definitely affect their core business, based on the paradigm of 'embodiment'. Moreover, scientists point to the fact that the way artists use technologies (typically in a different way than the scientists do) is sometimes inspiring for their own work and opens up new perspectives that were not visible in the context of a purely scientific practice.[4] The way artists work with certain technologies can be of interest to the scientists' practice as well. Finally, collaborations with artists may generate experiences that are valuable for later research projects that do not necessarily involve working alliances between artists and scientists. In all these constellations, interest and knowledge in certain technologies that are of importance in both artistic and scientific respects function as a common ground acting as a bridge between the different fields.[5] In these constellations, artistic practices and the artist's knowledge become involved in scientific argumenta-tion and give occasional impulses to scientific work.

In other scientific fields, such as life sciences and physics, a common ground concerning such tech-nologies is comparatively rare. In these contexts, artistic practices are more often perceived as a dif-ferent, quite opposite perspective, as a parallel world. Artists and their practices are seen as manifestations of another world, perhaps even of a widely opposed universe. The crucial question in such situations is how this 'otherness' is interpreted. Sometimes, the encounter with artistic views is described as a broad-ening of the horizon, which allows the scientists to see a familiar phenomenon in a different way. The encounter with the artist's work in the lab is not understood as concretely useful to the core busi-ness of science but rather as enriching for a personal perspective. The relationship between art and sci-ence is partly interpreted in a complementary way, which sometimes reprises old dichotomies. In these, scientific practices are typified as cognitive, rational, objective and highly specified and are distinguished from artistic practices, which are characterized as emotional, subjective, decorative and communica-tive. Two scientists, for example, characterized the difference between their perspective and artistic approaches by stating that the artist's work is pri-marily emotional, yet not necessarily 'irrational', in contrast to their own work, which they typified as 'rational'. One physicist emphasized that laboratories need people who are allowed to ask certain questions without seeming ridiculous when doing so. For physi-cists, a lot of questions that are emotional by nature or based on metaphysical dimensions are taboo, that is, a 'non-subject.' Artists are thus perceived as a possible resource for overcoming restrictions nec-essary to the specialized scientific work of physics. According to a number of sociological studies, artistic virtues have permeated broad parts of the capitalist working world, so that artists are seen as suitable role models, not least for managers (Boltanski & Chiapello 1999; Menger 2002). In the case of scientists, such an understanding of the artist as a role model is widely lacking. They prefer to typify artistic practices as *complementing* scientific approaches. What is especially remarkable is a perspective of the relationship between the arts and sciences that considers art as a medium that can communicate (even popularize) scientific work. This perspective is based on the assumption that science is distant from the public, that scientists live and work in an 'ivory tower', and that artists may act as a bridge. This vision appears every now and then in narratives of scientists; however, it is mainly present among program directors. So, in their study on different modes of interdisciplinary work, Barry, Born, and Weszkalnys (2008, p. 29) state that in such programs in the UK, art is often seen as playing a mediating role: 'One of the key justifications for funding art-science, particularly in the UK, has been the notion that the arts can provide a service to science, rendering it more popular or accessible to the lay public or publicizing and enhancing the aesthetic aspects of scientific imagery.' Artists are not only expected to mediate scientific and technological processes but sometimes even to 'humanize' them and to improve science and technology. In this case, the artists were ascribed with virtues that were usually associated with psychologists or healers. Their presence in labs is thus understood as a medium of control in a perceived 'runaway' technological culture (Leach 2005, p. 153).

To sum up, art is confronted by new expectations in the context of art-science collaborations and is associated with specific social functions. The perception of art as a field of application for scientific concepts is thus an important dimension, but not the only one. Artistic practices are also partly attested as having inspired scientific work. Other interpretations go further and ascribe a crucial social importance to the presence of artists-in-labs for the implications of scientific and technological developments.

Heterogeneous perceptions – interlocking dynamics

The actors' interpretations of the relationship between the arts and sciences are seldom completely congruent. Artists and scientists may not only differ in the forms of knowledge that they use but also in their interests and prospects. One could suppose that serious differences in the understanding of artistic and scientific work make interactions between artists and scientists difficult and collaborations almost impossible. Interestingly, this is only partly true. Different understandings of the natures of art and science are generally less problematic than one would expect. Such differences hardly prevent artists and scientists from collaborating and understanding each other in terms of more specific tasks. Besides the interactive co-presence that allows for ad hoc translations, it is of crucial importance to the mutual understanding process that most artists are not 'innocent' when they enter laboratories. In fact, some have remarkable experience in terms of new technologies, engineering, etc. Many artists that are interested in art-science collaborations have already had experience with scientific approaches. In addition to an education in the arts, some of them have an educational background in technology or science (they have studied mathematics, life sciences or engineering, for example), others have already participated in art-science programs or have been independently engaged in such working alliances. In addition, scientists are becoming more and more experienced with artistic work. Although additional educational backgrounds in the arts are rather rare among scientists, some who frequently work with artists gain considerable insight into artistic processes. Congruent world views and congruent conceptions of art and science relations may not be necessary for collaborative work. However, a wide mutual understanding and a far-reaching interest in each others working practices is a precondition for such collaborations to become long-term projects and independent working alliances. Working alliances are highly demanding with regards to finding a common ground, especially if they are not substantially backed by organizational patterns.

Art and science collaborations as instruments for cultural support create highly diverse constellations. The vision is certainly not appropriate that these programs lead to a 'synthesis' or a fusion of artistic and scientific practices (Barry, Born & Weszkalnys 2008, p. 22). Even the term collaboration, which is often used to refer to such encounters – as in this essay – is in a way misleading. It is euphemistic in the sense that it plays down the varied interactive forms and sometimes even opposing interpretative dynamics in the relationship between the arts and sciences. However, this constellation does not contradict the assumed importance of a 'fragmental type of differentiation' in contemporary society. Actually, the new combination of actors with heterogeneous backgrounds is likely to be accompanied by an experience of foreignness and an intensified need to explain individual positions and to define and draw distinctions. New institutional forms and combinations are not necessarily related to 'advanced' interpretations of the relationship between the arts and sciences. But it is highly probable that stereotypic interpretation patterns are challenged by the concrete and complex realities the encounters involve, if the residencies run long enough for a certain amount of mutual acculturation to occur. Even though the vision of synthesis and disappearing differences is hardly an apposite description of the processes that are concretely happening in labs, it has to be stated that art and science programs further the interlocking of artistic and scientific practices and boundary shifts on several levels. In combination with other dynamics, for example the constitution of research centers in the domain of electro-acoustics, the initiation of study programs at the intersection of engineering and art, or the foundation of small companies specialized in technology in the arts, art and science programs contribute to an intersectional domain and the formation of subjects with interdisciplinary profiles. Usually, the actors engaged in this domain clearly belong to a certain field in terms of their institutional position, and they have a corresponding reputation, for example as a physicist, composer, or choreographer. Typically, such a standing in a specific field is a precondition for participation in art and science programs, too. However, due to former associations or recurring collaborations, these actors have gained a broad knowledge of artistic and scientific problems and dispositions. Barry, Born, and Weszkalnys (2008, p. 40) have discovered that such dynamics are especially likely among university-based artists in

Australia and the USA who have been 'able to achieve intensive collaborations with scientist colleagues and prolonged encounters with scientific environments, thereby incorporating scientific problematics into their work [...]. Moreover such conditions provide the basis for transcending the disciplinary division of labour through a commitment to developing interdisciplinarity in one person.'

To gain a deeper understanding of the long-term consequences of the encounters' concrete impacts on artistic and scientific practices, the biographies of the actors, in the sense of educational and professional histories, should be studied further. What are the educational effects of the particular encounters? How do such effects stand in relation to the main professional preoccupations? Furthermore, the history of the objects and instruments developed during residencies should also be investigated more intensively. Which forms of knowledge and divisions of labor are they concretely based on? What happens when these instruments or objects leave the lab? In which contexts do they appear? Do they become border crossers or are they primarily received in the artistic field? This kind of research is crucial in order that social science may keep up with these ongoing cultural dynamics. However, such explorations are not only of scientific interest. A broader understanding of the dynamics in the intersection between arts and sciences is pivotal to learning more about the possibilities and restrictions of such cultural programs and working alliances. The encounters between artists and scientists that are currently taking place in contexts such as the artists-in-labs program have the potential to challenge existing concepts of authorship and productivity. Indeed, this potential is an opportunity rather than something that will inevitably realize itself. Thus the examination of the manifold dynamics may help to know more about the possibilities of 'integrating heterogeneity without losing innovative diversity' (Rammert 2003).

Notes

1 An example of the discussions examining art as science can be found at Kunst als Wissenschaft – Wissenschaft als Kunst, accessed 20 October 2009, <http://www.kunst-als-wissenschaft.de/de/index.html>

Interviews with Laurie Anderson on her NASA residency appeared, for instance, in the New York Times and Süddeutsche Zeitung, accessed 20 October 2009. <http://www.nytimes.com/2005/01/30/magazine/30QUESTIONS.html>. <http://www.sueddeutsche.de/kultur/560/408335/text>.

2 Hostettler, 2008/09, *Light Reaction – Dimensions of apparent invisibility* accessed 20 October 2009, <http://www.sylviahostettler.ch>.

3 A review of Ping Qiu's installations appeared in the Newsletter of the North American Benthological Society (NABS) 2009 'Benthology and artisitc expression – Part 1', in The NABS Newsletter, Issue 4, pp. 3–4, accessed 20 October 2009, <http://spinner.cofc.edu/~fwgna/downloads/NABSnewsletterIssue4.pdf>.

4 See, for example, Rolf Pfeifer's statement on the intelligence of the body and the use of robotics by artists, accessed 20 October 2009, <http://www.swissinfo.ch/08/flash/videoplayers/vp_standalone22.swf?sid=9978222>.

5 An interview with Jürg Gutknecht, Professor of Computer Science at the ETH Zurich, who has a broad range of experience in cooperation with artists in varied organizational constellations, was highly illuminative regarding the role of the computer in art & science collaborations and the question of how artistic practices are perceived by scientists.

References

Amann, K & Hirschauer, 1997, 'Die Befremdung der eigenen Kultur: Ein Programm', in St Hirschauer & K Amann (eds), *Die Befremdung der eigenen Kultur: Zur ethnographischen Herausforderung soziologischer Empirie,* Suhrkamp, Frankfurt/M, pp. 7–52.

Barry, A, Born, G & Weszkalnys, G 2008, 'Logics of Interdisciplinarity', *Economy and Society,* vol. 37, no. 1, pp. 20–49.

Baudelaire, 1976, 'Le peintre de la vie moderne' (1863), in C Baudelaire, *Œuvres complètes,* II, C Pichois (ed.), Gallimard, Paris, pp. 683–724.

Becker, HS 1974, 'Art as Collective Action', *American Sociological Review,* vol. 39, no. 6, pp. 767–776.

Behnke, Ch, Dziallas, Ch, Gerber, M & Seidel, S (eds) 2008, *Artist-in-Residence: Neue Modelle der Künstlerförderung,* Verlag für Wissenschaft und zeitgenössische Kunst an der Leuphana Universität Lüneburg, Lüneburg.

Benjamin, W 1991, 'Charles Baudelaire. Ein Lyriker im Zeitalter des Hochkapitalismus' (1940), in W Benjamin, *Abhandlungen, Gesammelte Schriften,* I/2, R Tiedemann & H Schweppenhäuser (eds), Suhrkamp, Frankfurt/M, pp. 509–690.

Boltanski, L, Chiapello, È 1999, *Le nouvel esprit du capitalism,* Gallimard, Paris.

Born, G 1995, *Rationalizing culture: IRCAM, Boulez, and the institutionalization of the musical avant-garde,* University of California Press, Berkeley.

Boulez, P 1986, 'Technology and the composer', in P Boulez, *Orientations: collected writings,* JJ Nattiez (ed.), Harvard University Press, Cambridge/MA, pp. 486–495.

Bourdieu, P 1996, *The rules of art: Genesis and structure of the literary field,* Polity Press, Cambridge.

Bydler, Ch 2004, *The global artworld Inc.: On the globalization of contemporary art,* Acta Universitatis Upsaliensis, Figura, Nova series, Uppsala.

Foster, H 1996, 'The artist as ethnographer', in H Foster, *The return of the real: The avant-garde at the end of the century,* MIT Press, Cambridge/MA, London, pp. 171–203.

Geertz, C 1993, *The interpretation of cultures: Selected essays,* new ed., Fontana Press, London, p.14.

Glauser, A 2009, *Verordnete Entgrenzung. Kulturpolitik, Artist-in-Residence-Programme und die Praxis der Kunst,* transcript, Bielefeld.

Graña, C 1964, *Bohemian versus bourgeois: French society and the french man of letters in the nineteenth century,* Basic Books, New York.

Greve, J & Heintz, B 2005, 'Die 'Entdeckung' der Weltgesellschaft: Entstehung und Grenzen der Weltgesellschaftstheorie', in B Heintz, R Münch & H Tyrell (eds), *Weltgesellschaft: Theoretische Zugänge und empirische Problemlagen,* Sonderheft der Zeitschrift für Soziologie, Lucius & Lucius, Stuttgart, pp. 89–119.

Heintz, B 1993, *Die Herrschaft der Regel: Zur Grundlagengeschichte des Computers,* Campus, Frankfurt/M.

Knorr-Cetina, K 1981, *The manufacture of knowledge: An essay on the constructivist and contextual nature of science,* Pergamon, Oxford.

Knorr-Cetina, K 1999, *Epistemic cultures: How the sciences make knowledge,* Harvard University Press, Cambridge/MA.

Kwon, M 2004, *One place after another: site-specific art and locational identity,* MIT Press, Cambridge/MA, London.

Latour, B & Woolgar, 1979, *Laboratory life: The social construction of scientific facts,* Sage, Beverly Hills.

Leach, J 2005, 'Being in Between': Art-science collaborations and a technological culture', *Social Analysis,* vol. 49, no. 1, pp.141–160

Menger, PM 2002, *Portrait de l'artiste en travailleur. Métamorphoses du capitalism,* Seuil, Paris.

Pollner, M & Emerson RM 2001, 'Ethnomethodology and Ethnography', in PA Atkinson, AJ Coffey, S Delamont, J Lofland & LH Lofland (eds), *Handbook of Ethnography,* Sage, London, pp. 118–135.

Rammert, W 2003, 'Zwei Paradoxien einer innovationsorientierten Wissenspolitik: Die Verknüpfung heterogenen und die Verwertung impliziten Wissens', *Soziale Welt,* vol. 54, no. 4, pp 483–508.

Stichweh, R 2000, *Weltgesellschaft: Soziologische Analysen,* Suhrkamp, Frankfurt/M.

ON THE NATURE OF INTERACTIONS Lloyd Anderson

Earthrise is the name given to NASA image AS8-14-2383, taken by astronaut William Anders during the Apollo 8 mission, the first manned voyage to the Moon. The photograph was taken from lunar orbit in December 1968 with a Hasselblad camera loaded with colour film. It is an image of the Earth as it appears from deep space, rising over the lunar horizon, and it was a view that no one had seen before. Hanging, half visible, in the black void of space was a beautiful, fragile-looking blue planet embraced by swirling white clouds. It was an image that changed our perception of where we lived.

Apollo 8 had entered lunar orbit on December 24, 1968. That evening, the three astronauts, Frank Borman, Jim Lovell and William Anders, made a live television broadcast from lunar orbit, showing pictures of the Earth and the Moon. 'The vast loneliness is awe-inspiring and it makes you realise just what you have back there on Earth', Jim Lovell said. They ended the broadcast saying 'For all the people on Earth the crew of Apollo 8 has a message we would like to send you', and they read the passage from the Book of Genesis where the light is divided from the darkness.

Earthrise was an image that was able to be captured because of the culmination of the most sophisticated technology in existence at the time and the most advanced scientific thinking in rocket propulsion, computing, electronics, and so on. This technology defined the USA's superiority at the leading edge of scientific discovery. It was a photograph of where we lived, taken at a time of great social change. We saw a very beautiful and delicate thing; something

fundamental and profound. In California, the love generation and flower power was at its height, with people trying to redefine the meaning of modern life; people 'tuning in and dropping out', people 'going back to the garden'. In Europe there were student riots and protests. And here was an image of Earth as a delicate living thing, to be nurtured and cared for. Not ruthlessly exploited and turned into a waste heap or parking lot. The image supported Lovelock's developing Gaia theory, where Earth is viewed as a single organism, in homeostasis; a complex system involving the biosphere, atmosphere, oceans, and soil; an optimal physical and chemical environment for life. It made us think again about where we lived.

As a child I had the poster of Earthrise hanging on the wall of my bedroom. I had pestered my parents to buy it for my birthday. To me it was a beautiful picture, a piece of visual art I wanted to look at each day, and perhaps to remind me of where and who I was. It didn't seem in the least bit 'sciency', but was an observation of nature, a statement of fact. It didn't need categorising and it was iconic.

In any discussion of science and art, sooner or later we have to grapple with categorizations not unlike dividing light from darkness. An artificial dichotomy has been constructed for which we can perhaps lay the blame at the door of C.P. Snow and his 1959 Rede Lecture on the *Two Cultures*, where he referred to 'the literary intellectual and the natural scientists, between whom there is a profound mutual suspicion and incomprehension'. As Rhonda Shearer and Stephen Jay Gould wrote in a 1999 essay on *Of Two Minds and One Nature*, 'Our propensity for thinking in dichotomies may lie deeply within human nature itself. Our tendency to parse complex nature into pairings of "us versus them" should not only be judged as false in our universe of shadings and continua, but also (and often) harmful, given another human propensity for judgment'. They believed 'the contingent and largely arbitrary nature of disciplinary boundaries has unfortunately been reinforced, and even made to seem "natural" by our drive to construct dichotomies – with science versus art as perhaps the most widely accepted of all. Moreover, given our tendencies to clannishness and parochiality, this false division becomes magnified as the two, largely non-communicating, sides then develop distinct cultural traditions that evoke mutual stereotyping and even ridicule'. They went on to say 'the worst and deepest stereotypes drive a particularly strong wedge between art and science. We do not, of course, deny the differences in subject matters and criteria (empirical versus aesthetic judgment) in these two realms of human achievement, but we do believe that the common ground of methods for mental creativity and innovation, and the pedagogic virtues of unified nurturing for all varieties of human creativity, should inspire collaboration for mutual reinforcement'.

Shearer and Gould then illustrated the potential junctions of art and science, referencing Renaissance figures in an earlier age that 'did not recognise our modern disciplinary boundaries and did not even possess a word for the enterprise now called science'. In more modern times, they pointed to the work of Marcel Duchamp. And there is a line of argument that now, at the beginning of the 21st century, these divisions are once again unnecessary because new technologies and global concerns have blurred the boundaries between what is an artistic and a scientific endeavour.

To my mind, this, too, is an artificial construct, an aspiration rather than a reality. It sends us in the wrong direction, believing that somehow there is only the creative process and all differentiation is unhelpful. Yes, new technologies mediate both science and art. But the work of the artist and the scientist is very different in their practical and fundamental nature. One is individualistic, unique, spontaneous and driven aesthetically. The other depends entirely on carefully controlled conditions, strict repeatability and an empirical base. Francis Bacon's studio looks nothing like the human gene sequencing lab at Hinxton Hall near Cambridge, and operates in a very different way. And yes, both require careful observation and creative insights and a result that is accessible to peers. But we should not pretend that these mind sets and skill sets are simply interchangeable. Both artists and scientists are specialists, and both are very necessary to the well-being and development of the society in which we all live. What we decide to call the products of the endeavours of these specialists doesn't really matter. It can be a mathematical equation or a scanning electron micrograph and it can be aesthetically or logically pleasing. It can be an acoustically perfect soundscape, a cubist interpretation of relativity or a sectioned cow. The important thing is that at some point one specialist modifies the environment, or the cultural milieu, that the other then operates in. That is the interaction. You don't need to create or recreate chimeras.

I should have said at the beginning that I am a botanist. And a few months ago I was sitting in a scientific conference listening to a talk about standardisation of photovoltaic device characterisation by surface analysis methods. While I could superficially understand what amorphous silicon thin films were, using a first year undergraduate-level knowledge of chemistry, physics and electronics, I was truthfully not very engaged by this talk because it held little meaning to me. I just couldn't get excited or absorbed by it. Even within my own specialism, I would prefer not to sit through a meeting about molecular techniques for DNA sequencing or mechanisms for ion transport through cell membranes. My real passion is at a different scale: landscapes, plant communities, and the dynamics of populations. And right then, in that conference, I would have been very happy to be walking across a hillside, and looking at the distribution of a plant species and trying to understand it in terms of the types of soil, the effects of herbivores, and the competition from other plant species. Sitting in that conference I felt I was in the wrong place.

The conference coffee break was in a big open space without windows, providing the time for socialising, or rather for networking as it is properly called; time for researchers to make new contacts and connections, for an exchange of knowledge and ideas, raising the possibility of new collaborations. But this open space was, invisibly, quite closed off. If you didn't work on sustainable energy technologies, it wouldn't be the best use of your time to be there. In fact, you wouldn't particularly want to be there because you would feel ignorant and excluded.

The day after the scientific conference was spent visiting hi-tech research institutes in a nearby city dedicated to research and innovation; a science city with immaculate avenues and pavements lined with trees and a still, quiet hush. We were taken to a large white building that looked like a cathedral dedicated to science and given a demonstration of a mobile harbour, where clever robotics engineering was used to keep a floating crane stationary. Even in very rough weather, it could lift containers off ships anchored out at sea. In a rather abstract way, this was followed by a visit to a Buddhist monastery, where we changed into robes and were instructed by the head monk in how to behave and how to bow, and given an explanation of the purpose of each of six ornate wooden temples, one by one. We then had to perform 108 bows, thread 108 wooden beads onto a string while listening to the monk chant, and have a meal of vegetables and rice while sitting cross-legged on the floor. Even to my cynical mind, I could see that this was a chance to find peace within oneself.

But these were other places, with their own sets of rules, instructions, words, and customs, where one felt one did not really belong. And these places contained people who had specialised, who had learnt their craft and were now using it in creative ways, be that researching solar cells, moving freight containers, or meditating in front of the Buddha. Each institution had a 'negotiated space', as the social scientists call them, where the specialists and the public could meet (quite literally on the periphery of the enterprise). But, one might add, the knowledge of the specialist was clearly a form of power.

'Agora' is the Greek word for a market place, and in ancient Greek city-states the market place was where people met and exchanged news. The agora marked the historical emergence of a public space which was 'neither controlled by the ruler nor relegated to the private sphere' (Nowotny et al. 2001). It was also the place where public announcements were made. And later in Europe one can imagine farmers and their labourers living a quite isolated life working the land, but once a week loading up their carts and taking their produce to the nearby village or town market, buying and selling but equally importantly meeting people and talking about the harvest or the health of their crops and animals; learning the news and gossip, and being part of the community.

If the agora was the market place to trade but also to exchange news, ideas, and knowledge with others, and to encourage many different people to come, then the natural philosophers and the artists would also have to make the time to be there. They would have to lock up their laboratories or studios and take themselves there in a frame of mind that was open to meeting new people, ready to contribute to unexpected discussions and perhaps, most importantly, to not think of themselves as superior to anyone else. (As an aside, that, effectively, is the idea behind the 'open space' methodology practised today by many facilitators at workshops designed to explore new areas, especially at the boundaries between disciplines.)

In *Re-thinking science: knowledge and the public in an age of uncertainty* (2001), Helga Nowotny, Peter Scott and Michael Gibbons view the agora as a social space where a transformation takes place, through 'the movement of contextualised knowledge into its context of implication (rather than the context of application)'. Here 'reliable knowledge (the hallmark of science) is superseded by socially robust knowledge'. This is an uncomfortable space for many scientists, who believe there is a set of 'inviolable principles, rules, methods and practices which are said to constitute the essence of science and cannot be discarded without endangering the whole enterprise'. The agora is not seen as an unstructured post-modern space; it describes a public space where 'science and society, the market and politics, co-mingle'.

Nowotny goes on to note that, historically, 'With the triumph of free-market capitalism and liberal democracy, this public space has transformed into an arena not only for market exchanges but also for open political discussion, an arena where criticism could be voiced openly, where public opinion was formed and political consensus reached' and adds that 'increasingly, the desires of both "consumers" and "citizens" were articulated in this public space'. This leads to consideration of the sanctity of the scientific enterprise. 'The argument (of Plato) about who has access to the realm of nature and the natural order and, therefore, who properly comprehends its laws, is still at the root of tensions that persist to this day. Science can no longer base its cognitive and social authority on a claim to have unique access to the order of the natural world (the understanding of which takes precedence over the understanding of the social world). Assertions about the autonomy of science do not carry much weight in the agora'. Crudely put, Nowotny believes that the walled-off realm of the scientist has to open up, and its sacrosanct knowledge become fair game for discussion and argument by others.

So we have the idea of a re-emerging Renaissance figure, where science and art are either fused in one person, or become irrelevant to that person, as in what John Brockman, the New York literary agent calls the *Third Culture* (Brockman 1995). Or where science and art interact in the medium of the agora, a space designated for direct negotiation and discussion between the different communities of interest. I could add to these two constructs – which essentially distinguish between art illustrating science and art interpreting science – the many schemes that have sought to fund interdisciplinary projects involving both artists and scientists. We can try to distinguish between illustration and interpretation, but all have in common the supposition that the interaction between artist and scientist must be a direct one. There is also a suggestion or hint that specialists should cede power to the broader-niche generalists.

It is useful, at this point, to borrow certain concepts from the world of ecology, in order to throw a different light on the interaction between scientists and artists, and to think of the agora as a setting for this interaction. There are two central concepts in ecology that have a bearing. The first concerns the kinds of interaction that occur within and between plant species. The second concept concerns the structure of plant communities, and the difference between generalist and specialist species. The agora here is perhaps akin to the broader, general environment rather than a specialist niche, and perhaps we all have

to operate in both (though sometimes we may feel out of our comfort zone).

If we turn first to mechanisms of interaction between species, a plant may influence its neighbours by changing its environment, for example by reducing light intensity, changing light quality, transpiring limited water, absorbing limiting nutrients, sheltering or excluding predators, or enriching the soil with organic matter (Harper, 1977). The changes may be by addition or subtraction. This means that interactions between plant species are mediated by the environment through the 'response and effect' principle (Goldberg & Werner 1983), which states that the plant and its environment modify one another so that the environment causes a response in plant function and growth, and the plant then has an effect upon the environment by changing one or more of its factors. Plant morphology and life history are governed by the environment, but at the same time the plant can change its environment. The nature of the interactions within and between species thus concern the ways in which a plant can influence its neighbours by changing its environment, directly, by addition or subtraction (of nutrients, for example), or indirectly (e.g. by encouraging insectivores) (Harper 1977).

There are a number of possible outcomes of the interactions between two species. Species A may affect the environment in a negative way for species B, or in a positive way, and hence an individual of species A may cause an increase (+), decrease (–), or have no effect (0) on the fitness of an individual of species B. The five resultant interactions (–,–), (–,0), (+,+), (+,0) and (+,–) have been defined in various

with digital colour enhancement, to make art out of science. Or, as a recent feature on BBC World put it while reporting on the Wellcome Trust Image Awards 2009 – an exhibition of stunning photographs of capillary networks, liver cells and summer plankton among other things – 'a fusion of art and science, science as art'. Here we are back to illustration.

Martin Kemp said of his book *Visualisations* (2000), that its focus was 'less in looking at the influence of science on art, or vice versa, but at shared motifs in the imaginative worlds of the artist and scientist'. For Kemp, 'too many of the increasingly fashionable art-science initiatives seemed to me to be operating at a surface level, in which obvious points of contact were simply narrated or in which objects from art and science were juxtaposed without really interpenetrating'. An organisation in the UK that is more interested in interpretation than illustration is Arts Catalyst, which 'brings together people across the art/science divide and beyond to explore science in its wider social, political and cultural contexts'. Their work is 'primarily invested in developing a dual discourse within the contemporary arts' and they 'produce provocative, playful, risk-taking projects to spark dynamic conversations about our changing world.' They have included many new and interesting artists, such as Simon Faithfull and Lise Autogena.

Turning to sci-art funding schemes, artistic and scientific administrators have, for some time, strongly believed that inter-disciplinarity is 'a good thing' and that creativity and innovation are to be found at the boundaries of disciplines. And so they have set up awards and incentives to encourage collaboration between scientists and artists. I don't think this was necessarily the result of a bottom-up demand from artists and scientists keen to work together, but rather a belief that enquiring and experimentally-minded people will create novel and interesting results. What actually happened was that the aspiring artists, seeking financial support, would propose an interesting collaboration – and some of them have been quite extraordinary and profound – and a slightly left-field scientist would lend their patronage. The result was, first and foremost, a piece of art, not a piece of science.

Thinking back to the sci-art consortium convened by the Wellcome Trust and to which the Arts Council, NESTA, British Council and the Calouste Gulbenkian Foundation contributed, there are some excellent examples, notably between siblings. *Medusae* was a collaboration between the artist Dorothy Cross and her brother Tom Cross, a marine biologist, 'to pursue investigations focusing on the aesthetic, anthropological and scientific aspects of jellyfish' (Warner 2003). *Primitive Streak* was a collaboration between the fashion designer Helen Storey and her sister Kate. As Helen puts it, 'Kate is a developmental biologist and she showed me her world. We had an idea to design a collection that communicated its wonder to others. Primitive Streak became a work that elucidated the first 1,000 hours of human life in textiles and dress'. Here were projects where art both illustrated and interpreted science, but primarily the former.

A more contemporary example of bringing art and science together, and interpretive in nature, was an exhibition recently shown at the Natural History Museum in London, curated by Bergit Arends. *After Darwin: Contemporary Expressions* showed nine artists' views of Darwin's book *The Expression of the Emotions in Man and Animals* (1872). Newly commissioned and pre-existing video, film, and writing from the author Mark Haddon, poet Ruth Padel, French photographer Gautier Deblonde, American artist Diana Thater, and video artist Bill Viola, among others, were brought together to explore Darwin's theory that expressing emotion is not unique to humans, but is shared with all animals. As the Guardian headlined its review of the exhibition, *Today's artists pick up where Darwin left off in mapping the expressions of humans and animals.*

The sci-art consortium, which Wellcome initiated, ran its course after a few years. It had funded a number of interesting projects both in an experimental phase and in production, but each partner wanted to develop his/her own ideas and schemes. Different interests were becoming stronger. Ken Arnold, Sian Ede, Bronac Ferran and Bergit Arends have all made important contributions since then. We, the British Council, hosted a series of informal evenings at the Union Club in Soho, London, inviting a range of people including visual artists, writers, musicians, conductors, physicists, mathematicians, graphic designers and museum curators to sit around a table with some wine and food and talk about the issue of scientists and artists working together on some form of new collaboration or application. It seemed to us that interdisciplinarity would result naturally and non-hierarchically in the context of an application, as Michael Gibbons had suggested in The New Production of Knowledge (1994). In fact, it didn't because

these were busy people with little time or spare energy to jump into the driving seat of an interesting side project.

Helga Nowotny, Gibbons' colleague, in *Re-thinking Science* believes that, 'Science has spoken, with growing urgency and conviction, to society for more than half a millennium. Not only has it determined technical processes, economic systems and social structures, it has also shaped our everyday experience of the world, our conscious thoughts and even our unconscious feelings. Science and modernity have become inseparable' (Nowotny et al. 2001). Bruno Latour, who Nowotny cites, argues that 'science and society cannot be separated; they depend on the same foundation. What has changed is their relationship' (Latour 1998).

So here we have the specialist, the scientists and science itself, modifying the cultural milieu in which we all live. And that modification of the environment is the important interaction with the artists and the arts, which requires the other to live in altered conditions. The artist, similarly, modifies the cultural milieu and so the thoughts and perceptions of everyone else. The agora, then, is simply the habitat or environment in which the community of specialists and generalists live. Perhaps it is the broader, fundamental niche of all rather than the narrower, realised niche of particular people, operating at a larger scale. Scientists and artists influence each other through their effects on the cultural milieu, the environment in temporal space that we find ourselves in. We don't need to find a name for the product of this interaction; how to categorise an image of *Earthrise*. And we don't need to force direct contact between specialists for there to be an interaction. Ideas will be assimilated and played back at altered wavelengths.

Damien Hirst, famously, cut a cow in half and pickled it in a display tank. This harked back to the dusty 18th and 19th Century collections in natural history museums and medical schools across many countries. Specimen jars on shelves of wooden cupboards with little neat handwritten labels saying 'longitudinal section', fixed in formaldehyde. Hirst appropriated this 'art' and turned it into something beautiful, interesting and disturbing. You could see a thin, outer layer of skin and hair above a layer of fat and then muscle, the bones, cavities and organs. Not neces- sarily remarkable to biology and medical students, but here, in a different space, asking questions about beauty and the grotesque, life and death, sentient animal and food. It drew you in, as a shrine to some- thing challenging and difficult. A question about what is art and what is science, and memories of juxtaposed colours and forms.

A.S. Byatt wrote a wonderful foreword to Sian Ede's book *Strange and Charmed* (2000), in which she says 'I believe the new images and understanding we are acquiring of the biology of consciousness will slowly change the forms of works of art in many disciplines'. She quotes Colin Blakemore, writing on the new problems to be addressed by the study of the brain, who says that neuroscience will 'undermine such cherished notions as spirituality, intuition and altruism – not by denying that people have them, but by providing rational accounts of them' (Blakemore, 2000). Byatt 'cannot believe that curiosity about the

science will not be more illuminating than automatic principled opposition to it. Just as understanding the complexity of genetic alterations will in the end surely produce... shifting metaphoric forms like Bernini's Daphne or Polke's witches and demons in caverns of poisonous and lovely pigments'.

Artists and scientists challenge us to re-think and re- imagine the scheme of things around us. They cast aside existing orders of the world and present us with new, different perspectives. They look at the world, record it, and play it back to us in an altered light, at a shifted wavelength. And scientists and artists are not divorced from the world or immune to it. To a greater or lesser extent they are influenced by the intellectual milieu and cultural context in which they live and work and which they, in turn, reshape.

References

Begon, M, Harper, JL & Townsend, CR 1986, Ecology: *individuals, populations and communities,* Blackwell Scientific Publications, Oxford.

Blakemore, C 2000, *The Independent on Sunday,* 2 January.

Brockman, J 1995, *The Third Culture: Beyond the Scientific Revolution,* Simon & Schuster, New York. See also <http://www.edge.org/3rd_culture>, accessed 8 March 2010.

Byatt, AS 2000, *Strange and Charmed,* (ed. Ede, S.). Calouste Gulbenkian Foundation, London, pp. 7–10.

Crawley, MJ 1986, *Plant Ecology,* Blackwell Scientific Publications, Oxford.

Gibbons, M, Limoges, C, Nowotny, H, Schwartzman, S, Scott, P & Trow, M 1994, *The New Production of Knowledge: the dynamics of science and research in contemporary societies,* Forskningsrådsnämnden, Stockholm.

Goldberg, DE & Werner, PA 1983, 'Equivalence of competitors in plant communities: a null hypothesis and a field experiment approach', *American Journal of Botany* 70, pp. 1098–1104.

Harper, JL 1977, *Population Biology of Plants,* Academic Press, London.

Kelly, K 1998, 'Essays on Science and Society: The Third Culture', *Science,* vol. 279, pp. 992–993. Also at <http://www.sciencemag.org/cgi/content/full/279/5353/992>, accessed 8 March 2010.

Kemp, M 2000, *Visualizations: The Nature book of art and science,* Oxford University Press, Oxford.

Latour, B 1998, 'From the world of science to the world of research?', *Science,* vol. 280, nr. 5361, pp. 208–9.

Lotka, AJ 1932, 'The growth of mixed populations: two species competing for a common food supply', *Journal of the Washington Academy of Sciences,* vol. 22, pp. 461–469.

Nowotny, H, Scott, P & Gibbons, M 2001, *Re-thinking science: knowledge and the public in an age of uncertainty,* Polity Press, Cambridge.

Schoener, TW 1988, 'Ecological interactions and biogeographical patterns', in Myers, AA & Giller, PS (eds), *Analytical biogeography: an integrated approach to the study of animal and plant distribution,* Chapman & Hall, London, pp. 255–297.

Shearer, RR & Gould, SJ 1999, 'Essays on Science and Society: Of Two Minds and One Nature', *Science,* vol. 286, no. 5442, pp. 1093–1094. See also <http://www.artscienceresearchlab.org/showcontent.php?contentID=23>

Snow, CP 1959, *The Two Cultures and the Scientific Revolution,* Cambridge University Press, New York.

Tilman, D 1982, *Resource Competition and Community Structure,* Princeton University Press, Princeton.

Vandermeer, J 1989, *The Ecology of Intercropping,* Cambridge University Press, Cambridge.

Warner, M 2003, 'Medusae', in Arends, B & Thackara, D (eds), *Experiment: conversations in art and science,* The Wellcome Trust, London.

BACTERIA, ROBOTS AND NETWORKS Dominik Landwehr

Today, the relationship between curiosity, imagination and networks at the interface of art and science are important concepts for many artists. In this paper, this dialogue will be examined through the work of some artists that have been funded by the Migros Culture Percentage – a unique organization in Switzerland.

The work place of Daniel Imboden is painstakingly tidy. Measuring instruments, a generator, and various projects – some under construction, others completed – are neatly lined up on shelves. Is this the work place of an engineer? It is not so easy to place Daniel Imboden in a specific category and it is not important to him whether his projects are considered art, engineering or entertainment. What counts instead are enthusiasm, engagement and also results, which can fascinate the audience time and again.

Daniel Imboden arrived at what he constructs today by many detours, but he still benefits from the fact that he was originally trained in the technical drawing of sanitary installations. A few years ago, with his accumulated knowledge of mechanics, material processing, electronics, and computer technology, he began to realize his dreams of building automats and robots. In his installation, entitled *Zeitraum (Time Space)*, he made a work where the hands of a clock play with a ball. His greatest challenge, however, has been to create robots that are more humanoid, and a series of prototypes in his workshop testify to the extent of these experiments. In a recent project he built a pair of music robots, called PETROL and SUPERMAX. The robot PETROL can drum its fingers rhythmically on its abdomen, which is constructed of an empty gasoline can. The SUPERMAX robot can even pluck a string that is stretched across its vacuum cleaner belly (Imboden 2007; Landwehr & Kuni 2008; Landwehr 2009).

Daniel Imboden is self-taught and thus experimentation forms the base of his working and learning principle: 'I build robots according to my own ideas. At first they often don't function at all the way I expected them to. In numerous trials and experiments I then begin to approach (something like) a functioning prototype.'

Through these inventions he is well connected across numerous networks. One of his closest friends is Flo Kaufmann from Solothurn, whose *bricolage universel* (Kaufmann 2009; Landwehr 2006) is related to some of Imboden's ideas. They were both guests of L'arc in Romainmôtier – a small guesthouse next to the medieval monastery, where artists and scientists are regularly invited on behalf of the Migros Culture percentage organization to exchange ideas and engage in undisturbed work (L'arc 2009).

It was not surprising that the head of the Artificial Intelligence Lab, Prof. Rolf Pfeifer, invited Imboden to give a presentation. 'His work is right in line with what we do at the lab', Pfeifer wrote at one point, 'there is no way around playing. Human beings gain knowledge on the basis of trial and error' (Pfeifer 2004, p. 97).

Differences and similarities
While there may be quite a few similarities in the methods used by the scientists in the Artificial Intelligence Lab and those used by the artist Daniel

Swiss mechanical engineer Daniel Imboden with one of his two music robots he built from scratch with used materials such as petrol cans or vacuum cleaners

Detail from Daniel Imboden's robot: The head can make some simple movements – remotely controlled by the master sitting behind his keyboard

Imboden, there are also many differences: Daniel Imboden also constructs robots in order to understand and isolate certain actions, such as pulling the string of an instrument or using a robot hand to drum on a metal object, but it is always difficult to ask an artist about his motivation and the content of his work. You do not always get a satisfactory answer and sometimes there is no answer, simply because the artist does not know the answer. Unlike a scientist, an artist is not obliged to give a reason for his work. Instead, the work may stand alone and does not have to illustrate a theoretical idea. If you ask Daniel Imboden why he makes robots he simply answers: 'I wanted to build a robot. Robots fascinate me and I would like to communicate this fascination.' He conveys a great sense of pleasure, which is derived from the playful aspects of his process.

The robots in the Artificial Intelligence Lab of Rolf Pfeifer may actually not look very different from those in the workshop of Daniel Imboden and the basic methods of trial and error are also used. But the scientists primarily use their creations to attempt to understand intelligence. They construct their robots because they want to comprehend the origins of very basic patterns of behaviour – the movements of a school of fish or how a child can discover its surroundings.

In fact, the scientists cannot investigate these questions entirely in a theoretical framework, but need to study them with the help of a three-dimensional model, which moves or is embodied inside a real environment. These models are often not objects constructed in a sophisticated and robust way, but are rather raw prototypes. They look more like objects assembled in an amateur workshop than in a design lab. The main objective of the scientist's efforts is not the performance of a robot in front of an audience. Instead, s/he wants to gain an understanding of how nature works. This understanding is eventually communicated to others by means of an article published in a scientific journal, because in the sciences, language is a major medium. It is used to communicate insights and also to develop them further, thus scientific discovery depends to a large extent on a text-based discourse. This is not so in the arts: the object – in this case a robot – does not require the help of any language because the communication process ends with the presentation of the robot to the public.

Despite these differences, there are similarities and even correlations between the scientists' and the artists' approaches and these provide an impulse for both to occasionally take an interest in each other. Nevertheless, while trying to find pointers or hints from their respective observations and discussions, they tend to keep their objectives of furthering the development of their own ideas and projects in mind.

Stefan Doepner's absurd automats

While play and the fascination with movement are two important elements in the works of Daniel Imboden, another artist might reveal entirely different expressions of robots and art. For example, the creations of Stefan Doepner (born in 1966 and founder of the Institute f18) constitute a different approach (f18 2009). Doepner constructed a painting robot that can move for hours back and forth across a large sheet of paper, periodically dipping its brush into a pot of paint. Little by little a random picture is created that constantly changes. In his newest installation 'Living Kitchen – Happy End of the 21st Century', a perfectly normal kitchen slowly develops a bizarre life of its own. Doors and drawers open and close, a radio turns on and off, a kitchen lamp flickers and might turn off.

Stefan Doepner and Daniel Imboden share many interests, but while Imboden attempts to create a more humanoid robot, Doepner's creations not only look different from animals or human beings, but also inspire different associations. Although their activity is target-oriented they are machines acting independently and autonomously, with goals that actually make no sense from a human perspective. In fact, Stefan Doepner's automats serve as a metaphor for technology gone astray, even for a technology that has become independent of human beings. Perhaps this interpretation of his humorous installation seems too one-dimensional, but Stefan Doepner's machines are first and foremost simply automats performing actions, which do not immediately reveal their purpose. His objects often provoke questions like: 'What is really happening here?' Or 'Why is this happening here?' Or 'Why is there no recognizable sense in the actions of these robots?'

As with other more provocative works of modern art, Doepner aims to raise questions or to irritate the observer rather than to provide answers. While scientists are also interested in raising vital questions they, unlike artists, always try to find the answers. Typically in science however, the answers may be tentative in character yet they are valued if they can lead to new questions.

The robots of Daniel Imboden and Stefan Doepner have some essential issues in common with the imagination and experience of the artists. The two artists differ in their backgrounds and paths of development. Daniel Imboden originally learnt the trade of a plumbing draftsman and taught himself the necessary mechanical and electronic-technical skills he needed for robotics, while Stefan Doepner studied art, but not mechanics or electronics. Neither of these artists have a scientific background and perhaps this is precisely what makes them interesting to scientists, as new ideas often develop when someone ventures into a completely different discipline. In this case, the scientist may be fascinated by the freshness and independence of a working artist who has little prior knowledge of the scientific field of Artificial intelligence but knows about the effect of artwork with analogue technology on the public.

The painting robot designed by Stefan Doepner. A robot, that can move back and forth periodically dipping his brush into a pot of paint, 2009

The Enigma cipher machine: The photo shows an Enigma K machine. This model was commercially sold and also used by the Swiss army in World War II

Digression: The mystery of the Enigma and the fascination of analogue technology

Certainly, analogue technology fascinates a broard range of the general public – and this fascination seems to grow as digital technology increasingly dominates our professional and ordinary lives. This is not only true for mechanical watches, which are still produced and sold today, but also for instruments that have long since become obsolete and can only be found exhibited in museums. One such object, for example, is called the Enigma cipher machine. It was developed before World War II but is still a popular reference for artists today. The German Enigma cipher machine was patented at the end of the 1920s and used commercially at this time, but was later used primarily in a military context (Landwehr 2008). The machine was easy to operate and promised unparalleled security, but this was an erroneous perception. As it turned out the Polish, and later the British code breakers – among them the great mathematician Alan Turing – managed to break the machine's encryption and even decipher German radio transmissions.

In many ways the history of the computer is connected to this machine and that is primarily due to the clever analysis of Alan Turing. In his theoretical article

On computable numbers, first published in 1936, he already proposed 'that some such machine would be capable of performing any conceivable mathematical computation if it were representable as an algorithm' (Hodges 1983).[1] It is an interesting coincidence that Alan Turing played an important role in Artificial Intelligence research and while working in Manchester after the war, he invented a test procedure to determine whether or not a machine could display intelligence. Today this test is called *The Turing Test.* The location of the secret code breaking operation was at Bletchley Park in England. Enigma machines were also built to automate the time-consuming first and foremost task of deciphering the operation of the so-called Bomb, an electro-magnetic machine that checked thousands of possible key combinations. A similar function was performed by Colossus, another machine constructed at Bletchley Park: unlike the Bomb, this machine no longer worked primarily with electro-magnetic elements such as relays, but rather with fast electronic tubes. For unclear reasons all machines and documentation of these operations were destroyed after the war and the fact that the code of the German Enigma had been deciphered was to remain a secret until 1974 when it became publicly known. In that year Capt. Fredrick Winterbotham – one of the code breakers – broke his silence and published a book on the secrets of the operation with the consent of the British government (Winterbotham 1974). Since then the story of the Enigma has been developed into a heroic tale, even a World War II myth, and every year newspaper articles, books and films appear dealing with some aspect of this story.

The Enigma was a relatively simple rotor-ciphering machine and in a time-consuming process every single letter could be encrypted. When looking at a picture of an Enigma machine with its top open, the principle quickly becomes very clear as one can see a kind of typewriter with a keyboard as input device and a board with lamps as output device. Located in between these two devices are the rotors, which actually perform the encryption work. When looking at an Enigma today, we are reminded of a clockwork mechanism, where we might easily understand how it works. This is one of the reasons why many people are fascinated by Enigma machines today. It was an ingenious machine: after all innumerable British, Polish and American specialists worked to break its code. But finally it only performed a simple operation – the encryption of letters. Unlike a computer its mode of operation is entirely analogue, its functions are visible and apparently comprehensible. Today, computers are different: they operate in a box and the mechanics are invisible. Certainly their actions are not always comprehensible and often specialists are the only ones who understand them.

Today, analogue technology and particularly analogue audio equipment such as record players and amplifiers still have their fans. For example, no lover of luxury watches would ever buy a digital quartz watch, as they believe watches have to be wound up manually. Automatic watches are a must for watch lovers; and the numerous additional functions such as phases of the moon or star constellations drive the prices of these mechanical works of wonder into astronomical spheres. Nobody is bothered by the fact that these watches never attain the level of precision of a modern quartz watch. On the contrary: This is an expression of the individual character of these particular analogue technical objects. Artists are also increasingly fascinated by old analogue technology. Does this also apply to Artificial Intelligence researchers who build simple models with many analogue elements in them? I think so. Is this pure coincidence? I think we are experiencing something like a renaissance in analogue technology. Perhaps this is also one reason for the rising interest in do-it-yourself technology.

An eye on bacteria

Migros Culture Percentage recently co-funded a project called: *Hackteria – Open Source Biological Arts*. The initiator of the project was Marc Dusseiller, a scientist who wrote his doctoral dissertation in the field of nanotechnology at the ETH in Zurich. For several years now, Marc Dusseiller has been involved in do-it-yourself electronics and is heavily engaged in the projects of the Swiss Mechatronic Art Society (SMAS 2009). This institution also offers workshops, where amazing technical gadgets and instruments are built using very simple electronics – for example a mini-synthesizer, whose sounds can be influenced by buttons and sensors but can never be entirely controlled. As Dusseiller himself suggested: 'Hackteria is a transdisciplinary project with the aim of develop-

A simple contraption including a cheap webcam – the device turns out to be a powerful microscope accessible through a personal computer (Photo: Marc Dusseiller, 2009)

ing and transmitting simple technologies for artistic expression in the realm of the live microcosm. The cultivation, interaction and observation of microscopic biological life forms shall be made accessible to a large group of media artists, nature explorers and musicians. The transfer of scientific methods into artists' workshops, home laboratories and public space in the Western world as well as in developing countries should lead to a democratisation of knowledge and raise questions such as: what do we eat? What lives in our environment? What opportunities does biotechnological research provide? Where are the border lines separating life and machine?' (Hackteria 2009).

The project builds on workshop experiences in the field of mechatronic art by converting cheap computer accessories, such as webcams for example, into instruments that can be used for one's own personal exploration of the world of cells and bacteria. But who should benefit from this experience: School kids in biology class? Curious adults? Artists? Dusseiller deliberately leaves these questions unanswered. Instead, he is 'interested in all different types of audiences'. The scientist turned artist also consciously avoids basing his project on any artistic insights or ideas. 'We provide instruments, we cannot anticipate ideas or projects'.

At first sight this idea appears strange: Why should an artist purposely refrain from creating his own works? Duseiller's definition of a work of art is unlike what we are used to: His work is no longer a specific and singular object but rather a particular situation or context or platform, in which art can be created by many. This concept can also be applied to Stefan Doepner's painting robot projects: Here the work of art is not primarily the sheet of paper covered with lines, although this is also possible and conceivable, but rather the entire complex situation of a robot applying paint to paper in random actions. Marc Duseiller's platform can also be seen as an arrangement intended to provide the conditions in which participants can produce art. Other examples of this contextual attitude can also be found in art history – such as the Dadaists' events in the 1920s or the Fluxus actions in the 1960s and 1970s. Although these two art movements are primarily seen as protests against the dominant and ritualized art of the time, they also offered creative people very individual reactive potentials, which by far exceeded some elements of protest.

The idea of platform provision is widespread in the media arts. For example, Austrian artist Gebhard Sengmüller has produced numerous media contrap-

Notes

1 Out of the many studies about the life and work of Alan Turing, the outstanding Biography from Andrew Hodges might be mentioned here: Hodges, A 1983, Alan Turing: the enigma, Simon and Schuster, New York.

References

f18 institute for art, information and technology, accessed 2 November 2009, <http://www.f18institut.org>.

hackteria, accessed 2 November 2009, <http://hackteria.org>.

Hodges, A 1984, *Alan Turing: the enigma,* Simon and Schuster, New York.

Imboden, D 2007, 'Supermax & Petrol', *dim-tech: daniel imboden – technische lösungen im bereich industrie und kunst,* accessed 2 November 2009, <http://www.dim-tech.ch/roboter.html>.

Kaufmann, F, Experimentelle Klanggeneratoren mit CMOS-Chips. pp. 72–83.

Kaufmann, F 2009, *bricolage universel,* Solothurn-suisse, accessed 2 November 2009, <http://www.floka.com>.

Landwehr, D (ed.) 2006, *Home made sound electronics: Hardware Hacking und andere Techniken: mit Andres Bosshard, Nicolas Collins, Verena Kuni, Norbert Möslang und Bruno Spoerri,* Merian, Basel.

Landwehr D 2008, *Mythos Enigma: die Chiffriermaschine als Sammler- und Medienobjekt,* transcript, Bielefeld.

Landwehr, D, Kuni, V 2008, 'Do-it-yourself noise generators and videomachines', *Home made electronic arts,* D Landwehr et al (eds), Merian, Basel, pp. 58–65.

Landwehr, D (ed.) 2009, *Werkbeiträge Digitale Kultur 2, Migros-Kulturprozent,* Merian, Basel, pp.14–17.

L'arc: Littérature et atelier de réflexion contemporaine, Migros-Kulturprozent Romainmotier, accessed 2 November 2009, <http://www.l-arc.ch/>.

Mediengruppe Bitnik, accessed 2 November 2009, <http://www.bitnik.org>.

Pfeifer, R 2004, Roboter bauen – ein Kinderspiel. oder: Probieren geht über Studieren, in: Dominik Landwehr: Playground Robotics. Das Hamburger Robotik-Kunst-Institut f18 und seine Schweizer Freunde. Merian, Basel, p. 97.

Science et Cité: wissenschaft und gesellschaft im dialog, accessed 22 November 2009, <http://www.science-et-cite.ch/de.aspx>.

Sengmüller, G 2000, *vinylvideo,* accessed 2 November 2009, <http://www.gebseng.com>.

SMAS: Swiss Mechatronic Art Society, accessed 2 November 2009, <http://www.mechatronicart.ch>.

Winterbotham, F W 1974, *The ultra secret,* Weidenfeld and Nicolson, London.

Wirth, N 2008, 'Als Computer noch erklärbar waren', *Neue Zürcher Zeitung,* 31. Oktober, accessed 2 November 2009, <www.nzz.ch/nachrichten/medien/als_computer_noch_erklaerbar_waren_1.1196876.html>.

ARTISTS-WHO-CARE! SHARED PERSPECTIVES ON SOCIAL AND ETHICAL RESPONSIBILITY Jill Scott

'Now only the deeply committed are active. The "me" generation has never stepped up to the plate and us old folks are getting tired' (Lippard 2008).

In the 60s during the Vietnam war, it was not very hard to drag artists out of their studios onto the streets nor find them in the role of 'political story-teller' for the public. These artists, who are now part of my generation, still fondly talk about their respon-sibility to engage with the public and how this focus drew them together as an art community. In this essay, I would like to outline how, today, the art/sci margin might become a healthy place to re-share controversy and discuss action, rather than a dangerous one for artists to be associated with. Moreover, there are a number of reasons why the sharing of this place with scientists might be beneficial. My approach is an 'out of the box' reflection about how to cultivate more responsible transdisciplinary teams on a practical as well as a theoretical level. This approach may require the roles of the fabricators and the discoverers in our society to be re-thought. Because this is such a large subject, I will attempt to address current social and ethical issues by using examples from the following debates; access to technological interactivity, caring about impairment and human neural perception, the impact of biotechnology on society and the challeng-ing task of adapting to climate change.

I will also expand on three major relevant claims, the first being that art can no longer deny the state of the very world in which it exists, and that when art has the freedom to be a creative, reflective process with the aim to produce a shift in perception for others, then all forms of the arts (sound, film, architecture, design, theatre, literature etc) that attempt to comment upon current ethical and social issues must also be recognized as art. Second, there are a growing number of contemporary art groups, which are more deeply committed and actively concerned about their inclusion in society. The artists who associate themselves with these groups are often specifically interested in the social impacts of scientific discovery and the ethical discourses surrounding the scientific process. And finally, when these creative results are embedded with a serious interest in science then the general public can shift their inter-relational equation between the sciences, the arts and the environment in which they live. The above claims not only require the formation of organizations, which foster transdisciplinary re-search or provide context provision, like the case studies featured in this book, but a restructuring of societies' attitudes toward the role of art in culture generally and the general publics' perception of scientific discovery.

Within the last decade, scientists themselves have begun to question their own career stereotypes and scrutinize the ethical boundaries of commercialism, since their respective industries can monopolize their own discoveries and larger stakeholders may even control the output of their results! The resultant debates about technology, human health and the environment are slowly becoming more shared with each other through the dissemination of knowledge. Society, with its growing literacy, seems to want sci-ence to 'come out' of its box and does not want to feel marginalized any more from the life sciences, physics, computing, or engineering.

Claim Three

By the same token the general public also needs creative, provocative and symbolic comments and interpretations from artists about the critical contexts of scientific discovery, in order to have some more solid meat for deeper reflection. Our histories are full of instances where committed groups like Greenpeace have proven that they are responsible enough to take actions into their own hands, but it is refreshing to find groups of scientists who feel the same way. In 2009, ENSSER or The European Network of Scientists for Social and Environmental Responsibility was formed to assess the applications of existing and future science and technologies. Here a group of concerned scientists assess the ecological health and socioeconomic impacts of scientific discovery and consider alternative opinions to mainstream science and technologies with the aim to strengthen citizens' interests over vested and corporate interests. As biologist and founder Angelika Hilbeck states, 'we now want the public to move beyond the mode of disinterested contemplation to something that is more participatory and engaged' (2010). As an artistic member of this same organization I was interested to find that some members would like to collaborate while others totally shun the experience. So how do we help both scientists and artists to learn to trust each other in the first place? Surely a 'hands on' experience of each other's environments and processes can only help to deepen the respect for each other's disciplines and work together to raise public awareness.

Citizens are even taking some responsibility into their own hands. Recently, in a project called Carrot MOB in Germany[3], citizens were asked via internet to favour re-sellers who were committed to spending their resultant profits on greening their own energy consumption. For a society that has always lived in the present tense of haves and have nots, full of those who want rather than those who need, these types of behavioural attitude changes prove that green activism can work in a positive way. If artists and scientists were to become more involved in local strategies like these, current social and environmental solutions for sustainability might be encouraged. As can be ascertained from the above paragraphs, the structural relationship between science, society and environment needs to be re-thought.

For purposes of clarification I have chosen four scientific discourses and relevant examples in the arts to illustrate this relationship, in other words, just a few instances of many correlations of responsibility. I also chose them because I have some personal experience as an artist in each context. In the first category, media arts, the scientific developments of technology and human computer interaction have divided the respective communities into the helpers and the profiteers. Second, I have chosen the field of neuroscience and the growing levels of its impact on the arts because shared potentials in sensory perception and attempts to understand human cognition can help us to understand ourselves and the way we think and behave. The third category biotechnology and its provocative counterpoint, bio-art, focuses on the impact these discoveries have on our society. Finally, I attempt to unpack our responsibility towards our environment and what some environmental artists are doing to address sustainable outcomes.

Dark Sky: Viewer interaction. Tiffany Holmes. Museum of Modern Art in Chicago

Interactive Startlight Result: *Dark Sky.* Tiffany Holmes. Museum of Modern Art in Chicago

Interactions with Technology

In a book entitled, *Art Science and Creativity – The Post Google Generation,* scientist and author David Edwards concludes that special laboratories must be set up for the youth of society (2009). So he set up *Le Laboratoire* in Paris as an alternative to 'our disciplinary based institutional crisis, because transdisciplinary changes need high tech institutions, which in turn can become clever instruments for social empowerment.' Although supported by arguments that innovation is a messy process, hundreds of French kids attend the labs events and are seen as the future inventors of interactive technology and (HCI) Human Computer Interface Design. Alternatively, the older interactive media are divided into the researchers who design technology to help people and attempt to humanize it for society and the profiteers who are mostly interested in the businesses of promoting the latest gadgets and participation in the internet to make money. The former group think that new technologies can be used to promote a better understanding of social issues or help particular groups who are excluded from access. In a project entitled *Constant Travellers* by architect Monika Codourey, she appropriated a set of methodologies from HCI analysis in psychology (University of Basel), in order to design a media artwork with accompanying software potentials of interactive

learning and non-linear narrative. After interviewing travellers about their mobility patterns, it became apparent that she could help them emotionally by designing an application for their mobile phones. This software could allow them to connect and trade travel information in-situ. Interestingly enough an architect thought about social connectivity as an antidote to loneliness in the context of a cold, isolating and even boring environment like an airport. By studying the problems of a mobile lifestyle, before considering building an application, she redirected innovation along a more caring pathway. In such contexts, the ethical issues of privacy of use and surveillance could be similarly considered.

Another media artist, Tiffany Holmes, took on a bigger challenge and thought about how interactive media could actually encourage the general public to learn something about energy conservation. She began by abstracting real scientific data about energy consummation and constructed animated eco-visualizations. For example, in *Dark Sky,* an installation presented at the Museum of Modern Art in Chicago, when viewers turn off electric bulbs, increasing amounts of stars appear on a nearby screen. The numbers of stars are relevant to the wattage saved with each action by each viewer. While in computer science, Human Computer Interaction is seen as a fledgling new business mostly relegated to ACM bank machines and graphical interfaces, the types of dynamic feedback used by Holmes puts the responsibility for conservation back onto the users themselves. Margaret Tan, an artist from Singapore, spent her residency in a very commercial micro engineering lab (CSEM), learning about micro technologies. However, she appropriated these techniques in order to construct a networked interface that might help foreign domestic workers in Singapore stay in touch with each other. Her colleagues, the scientific researchers were duly impressed by this endeavour because they would not have thought of such an application, nor would it ever be funded because the 'user group' was far too small and poor to buy it! Once my own transdiciplinary team encountered a similar reaction in a project called *eskin* that was designed for visually impaired people, in order to help them cope with a visually dominant world. As all of the above examples indicate, artists are interested in addressing the fact that in the engineering and computing industries the needy in society are not being prioritized. As the media theorist Paul Dourish recommended, technology still needs to be humanized in order to help people in our society (2001).

Furthermore, when artists and scientists collaborate on innovative potentials they can combine their processes of experimentation with education. As Adam Greenfield and a team of transdisciplinary practitioners have proved, augmented reality with digital technology can not only enhance the learning curves of underprivileged viewers but any 'hands on' interaction can also improve visual reasoning generally (Greenfield 2006). Interactive potentials precede,

but are also germane to all scientific and artistic creation, and are inherent in children. Therefore rather than be so profit driven and monopolized by companies[4], electronic technology and the internet must be used to transform the way we organize and seek knowledge. As has been shown above, some media artists[5] have started to combine these attributes with pertinent contexts from the social sciences and responsible content has been the result.

Neuromedia

But how can artists reach out to people with HCI technology if they do not know how mediated information is actually perceived by the brain? Scientists who study neurobiology or cognitive science attempt to understand how the body and the mind function by conducting medical excavations, in vitro experiments and perceptual analysis. From the mechanistic perspective of neuroscience, the nervous system is regarded as a highly sophisticated tool, which enables animals, including humans, to perceive the outer world and interact with the environment. For years artists have been experimenting with the fact that the very same object can be perceived in many different ways and cognitive scientists attempt to understand how the mind can make metaphorical leaps, take risks or offer subjective associations. Therefore, the interplay of media, art and neuroscience is a highly valuable transaction. For example, artists have already experimented with neuroscience in the following arenas – visual impairment, attention deficit, and bodily illusions.

In art, the hegemony of the eye is very strong in our culture, and in order to challenge this commitment

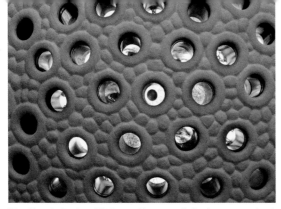

The Electric Retina. Jill Scott. Close up of the 'cones occulars'. The Brain Fair, Zurich, 2008

The Electric Retina. Jill Scott. Changes through observation. The Brain Fair, Zurich, 2008

to its ocular-centric, or vision-centered aesthetic, I spent many hours studying with neurobiologists[6] who were researching different types of visual impairment on the retinal phenotype of the zebra fish. The sensory processing perception of vision is subjective and by no means universal, and impairment causes many shifts in subjective viewing. In collaboration with the scientists we created *The Electric Retina,* a sculpture that combines retinal research with interactive media art. The aim was to gain a deeper insight into the genetic control of visual system development and function as well as the rod and cone pattern array of photoreceptors in the human retina. While the 'cones' display histological evidence (behaviour tests, molecular staining, cellular research images), the other side of the sculpture projects movies of fiction, shot from the perspective of the impaired subject, which shows how impairment might affect neural behaviour. Touch was used as an interface for the viewer to explore the sculpture, because in neuroscience, the effort is to not isolate vision from the other sensors; and to understand how activity-dependent plasticity works in the brain has been an important research track for a long time.[7] Similarly, artists and designers hope that by understanding various levels of cross modal interaction, the communication potentials of their HCI work might be expanded.

Since the 60s avant-garde composers have been interested in the potentials of how EEG sound patterns in the brain could add to the experience of bodily awareness or proprioception (e.g. Alvin Lucier 1965). The ability to efficiently integrate sensory information arriving from multiple modalities and from different spatial compartments is crucial in localizing ourselves and for navigating in the environment. As Italian Swiss artist Luca Forcucci discovered recently at the Laboratory of Cognitive Neuroscience at EPF in Lausanne, perceptual and cross-modal deficits in brain damaged patients may lead to pathological illusions and distortions of self-consciousness, such as out-of-body experiences. Such illusions may alter our awareness being localized in a given environment, and our capacity to self-identify with our own voice (Blanke 2010). The scientists attempt to understand this awareness of our bodies in a given environment by using Virtual Reality systems and brain wave analysis (EEG). Distributing sound to specific locations may help these people. In a collaborative project with the labs director Dr. Olaf Blanke, Luca Forcucci constructed the installation KINETISM, which explored the ways that sounds in the urban environment could be mixed with internal body sounds, so that the viewer can experience these associations. As new discoveries in neuroscience have proved, bodily action comes from multisensory integration and this information can be important for 'an artist who cares'. Even in his 1998 notes about interactive art at the Exploratorium, Peter Richards claimed that artists who share a deep-seated joy of learning can construct approaches that may present the public with new discoveries and understandings about their bodies in space (Richards 1998). Thus HCI can celebrate one of the most human of activities, the process of drawing meaning from embodied levels of attention.

Recently, New York artist, Ellen Levy closely examined the cognitive problem of Attention Deficit Hyperactivity Disorder (ADHD or AD/HD), and she then constructed an animation with Michael E. Goldberg, Director of the Mahoney Center for Brain and Behavior at Columbia University, that caused many viewers to confront the limits of their own perception. She stretched the attentional system of viewers in an art gallery context by having them experience their own propensity for inattention blindness. Viewers were tested by watching an animation, to see if overlaid images of a fast card game could distract them from noticing the slow disappearance of stolen museum artefacts. In doing so she posited the question: Would it be possible that the general public and its encounter with arts can become subjects for the gathering of scientific evidence? Opinions regarding ADHD range from not believing it exists at all, to believing there are genetic and physiological bases for the condition, as well as disagreement about the use of stimulant medications in treatment. While ADHD or AD/HD is defined as the co-existence of attention problems and hyperactivity in children,

with each behaviour occurring infrequently alone (Steven P. Wise and Robert Desimone, 1988), it is one of the most controversial diagnoses in neuroscience. It is not only important that Levy has addressed the topic of the adjudication of disease versus normality, but that by using the attention-based medium of art itself; she has also extended its boundaries.

The above artists[8] have been engaged in research with neuroscientists and they have offered interpretations of particular neurological states in the context of Neuromedia. These states included attention blindness, cross-modal perception, brain-wave analysis, and neural pathways, but there are many more perceptual inspirations to be found in neuroscience research. Indeed it seems that if artists can learn more about how human perception is affected by the environment they inhabit as well as by genetics, disease, and degeneration, they will understand more about our behaviour and how to shift the way we think! For in contrast to what most people think, we actually can retrain our brains and thereby change our bad habits in the future (Doidge, 2007)!

Provocative Bio-Art

In 1939, Alexander Fleming also attempted to shift his audiences perception when he presented his 'microbe paintings' at the Second International Congress of Microbiology in London. While the scientists in the audience largely ignored these strange paintings produced with pigmented bacteria, Fleming's discovery of penicillin later revolutionized medicine. In 2006, I attended a workshop on cellular and molecular biology with 20 other artists (held at Kings College in London) organized by Arts Catalyst. All of us shared a

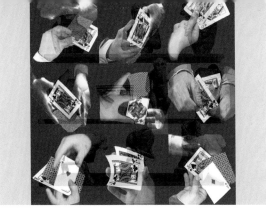

Stealing Attention: Ellen Levy, Looped Animation in which the Three-Card Monte is superimposed over antiquities looted from the Iraq National Museum. When the Queen of Hearts appears, an antiquity disappears, 2009

Art and art history students watching the animation, *Stealing Attention*: Ellen Levy at the Michael Steinberg Fine Art, New York, March/April 2009

fascination about other ways in which scientific truth could be captured but wondered how we could possibly compete with those luscious colour images from the cellular and the molecular world. Could they be integrated into our art practice? During the workshop we learnt to extract and pattern DNA, cross genes, engineer tissue and grow cells in-vitro, but unlike science students, we paused for ethical discussions about the controversies of biotech manipulation and their related industrial patents. As similar encounters in other countries have proven, artists who have been exposed to lab techniques or even been allowed to closely shadow the processes of scientific discoveries take on the challenge to be 'provocative' about ethics and biology.[9] (i.e. see SymbioticaA at the University of WA, Australia). This challenge is to make artworks or live events, which can possibly nurture bio-controversies for a broader public discourse.

In 2007, performance artists Hina Strüver and Matthias Wüthrich shadowed the scientific process of physically inserting DNA particles into the nucleus of the seed of a plant fired from a gene gun in order to create a genetically modified organism (GMO). Consequently, they created *Regrowing Eden* a three-part performance project with a related website <http://regrowingeden.ch> that attempted to raise public awareness about the growing of genetically modified agriculture in Switzerland, Brazil and Vietnam. While the performative installations functioned as poetic metaphors in different cultural contexts by constructing huge plant like structures, the website allowed a broader audience to create simulated 3D forms of GMO plants by themselves. The artists came to realize that scientific 'cultural hegemony' was truly separated from life on the streets and sought a variety of reactions from different countries. According to senior scientist Dr. Angelika Hilbeck at the Institute for Integrated Biology, ETH Zurich, 'we also had lots of discussions among us about how scientific discovery cannot stay inside the circles of the science community and be isolated from social political problems'.

As I mentioned earlier, this is exactly why the formation of groups like ENSSER is so important inside this circle (<http://www.ensser.org>). Every GMO scientist knows the story of Dr. Arpad Pusztai who fed rats with GM potatoes and claimed they caused lesions in the spleen (Ewen and Pusztai 1999) or the story of the farmer Percy Schmeiser who was 'caught' with GMO plants in his bio-granola that had been contaminated by pollen from another farm, which just happened to float over the fence (<www.peryschmeiser.com>), or the problems with Monsanto[10] and their lack of understanding about the need for seed diversity. When artists, who also care, make new interpretations about the risks of transgenic production in the developing world, they share their discourses with these scientists; they also come to the realization that it is high time for the responsibilities of biotech companies to be reviewed.

For the last decade artists have been adding to these provocations by taking on the manipulation of live materials and animals used inside the lab itself. For example, tissue engineering is promoted as natural, biologically based approach to repairing or replacing bodily tissue functions, but is actually based on a process of guiding bio-cultured tissue cells to grow on artificial polymer scaffolds. In 1996 artists Ionat Zurr and Oron Catts founded the Tissue Culture and Art Project, and one of their works was *The Pig Wings project* (2000–2002), constructed by growing pig mesenchymal cells (bone marrow stem cells) on bio-degradable/bioabsorbable polymers (PGA. P4HB) in the shapes of wings. Besides the obvious provocative implications of the 'flying pig' metaphor about the future of regenerative medicine, these wings constitute a seminal expression of the new bio-art relationship between live lab materials, anatomy and meaning. By engaging with a work like this, viewers soon realize that cells contain the matrix of code for most types of cell differentiation, but through the metaphor of 'flying pigs' they realize that cells could also become the ultimate morphing material. These artists have continued to cause public reflection by combining essential fabrication with poetic metaphor and questioning the realities of nature. Therefore, without the

interference of religious morals, scare tactics or dissent, they have married 'fabrication' with 'metaphor' in order to open up public debate about the future of bio-medical applications in the museum context.

In an alternative to the sculptural solution, experimental science writers like Donna Haraway highlight the ethical debates about transgenic animals and organs by publishing their analysis of 'bio factual' entities. For example, Onocomouse is a mutant mouse born with the breast cancer gene, a creature with no rights to existence outside the laboratory station itself (Haraway 1997) but a creature which alongside many other hundreds of lab animals (including yeast and bacteria) has been patented. In biotechnology the ethical debate ranges from pain induced experimentation on mice, to the potentials of rescuing genetic deficiencies in them by breeding mutants and attempting to revert their processes of degeneration. While these experiments are often related to human disease, observations of wild types in their natural habitats are paramount for comparisons. Brandon Ballengee, an activist eco-artist who works more in the field than in the lab, collaborates with volunteers to locate his specimens. By doing so he has found many examples of mutant growth in the limb buds of amphibians. And under the guidance of another scientist, Stanley Sessions, he eventually proved that these deformities could be attributed to an increase in 'selective predation' by dragonfly nymphs.[11] He shares these results locally, with under-privileged urban groups or suburban families; these local communities get involved in scientific discovery. Both Ballangee, through his community art/science works and Haraway, through her experimental sci-

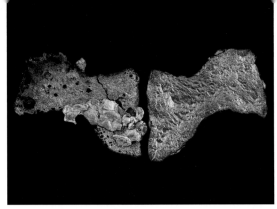

Untitled. 2000–2001, *The Pig Wings* project from The Tissue Culture & Art Project. Dimension: 4cm x 2cm x0.5cm. Others versions include, The Chiropteran Version, The Aves Version and the Pterosaurs Version

The Pig Wings installation: The Tissue Culture & Art Project as part of *conVerge*, Adelaide Biennale of Australian Arts, Art Gallery of South Australia, 2002

ence writing provokes discourses in society about the treatment of animals in the labs and the value of their observation in the field.

As the above provocative strategies by performers, bio-artists and writers show[12], working with the wet specimen of the scientific process or creating platforms for discussion all help to create symbolic metaphors that can in turn stimulate the public to engage in the discourses of biotechnology. Perhaps future creations could also help the biotechnological companies reflect upon their own value for society.

In science, the conventional accounts that biotechnologists offer about their successes to each other are certainly not value-free. As science writer Evelyn Fox Keller once suggested, 'The very language, tacit presuppositions, expectations, and assumptions shared by scientific researchers are very value-laden'. (2000). While the scientist's quest for knowledge in itself reflects the existence of a value system, biotechnology tends to entirely exist and finance itself within a political and economic climate that is full of values related to political priorities and capitalist interests. So far the art researchers who enter the gates of science are more interested in how to offer another angle about these foci or even invent new strategies.

Environmentally Active Art

Art researcher and artist Amy Lipton once coined the term 'ecovention' by constructing it from 'eco' and 'invention', labelling it as a strategy that artists use to attempt to transform ecologies. But the definition also incorporates art projects that employ an inventive strategy to physically restore a local ecology (Lipton 2002). In the post-modern art school education of the 60s and 70s, we were often taught that nature is a socially constructed idea, a scenario that now may or may not provide us with a considerable hope to re-build a sustainable future. We also learnt that the marketing of nature impinges on all factors of social reality and tends to construct nature as one that contains a certain set of human ideals and our relational place inside them. The general public also tends to treat the environment as 'a given situation' and so an issue like climate change can shake their very perception of 'the real'. Ecoventionists, it seems, care about what society might need to do in order to inhabit this planet in the future, while evolutionary artists wonder if nature should 'not just go ahead and evolve without us!' Greenmuseum.org took the first option and started to gather artists who want to improve society's relationship with the natural world. They also decided that because environmental art is 'ephemeral (or made to disappear or transform)' and 'designed for a particular place (and can't be moved)' or 'involves collaborations between artists and others, such as scientists, educators or community groups (distributed ownership)' that these variables can make exhibiting this work very difficult for traditional museums <http://greenmuseum.org/what_is_ea.php>. Therefore, once again, the internet offers the potential for these artists to band together for the support of sustainable ideals. However, the general public is still ostrasized from scientific facts about the very same issues.

For example in climate science the visual representations of past, present, and future predictions about climate change are often encompassed into models for specialists. These are divided into a relational network of the statisticians who collect the data; the modellers who place it into a simulated form and the users who use these models to make comparisons. The last people to receive the data are the politicians. In a recent workshop I attended on Climate Change at the Swiss Federal Institute of Technology (ETHZ), perhaps outsiders could sign up for the workshop, because these scientists are becoming very worried about the public perception of their fact finding missions. Recently, it became apparent that some climate scientists were being paid by large companies with vested interests to distort climate change evidence in the public realm. I claim that artists can, and should, not only work together with scientists to dispel these untruths but also with communities so that the scientific facts can be better understood.

When sound artist Andrea Polli, was in Antarctica for three months, she talked with many climate scientists about the affects of simulation models compared to the realities of data collected on the ground – an activity they aptly called 'ground truth'. In a resultant project called *Hello, Weather!,* a set of local weather network stations have been installed in different local communities around the world. These citizen weather stations are designed to demystify the collection of weather data and seven professional weather stations are now in operation. In addition to the Zurich station, two long-term stations are currently in operation in New York City, one at the Eyebeam Art and Technology Centre and one in Long Island City. There is a station at the Centre for Contemporary Art in Santa Fe, New Mexico, and one at the Audubon Centre in Los Angeles. This work builds on the existing international phenomenon of Personal Weather Stations in which enthusiasts worldwide combine DIY technology with organized web forums for collecting and analyzing data. By allowing them to use this technology in an easy way that is normally inaccessible to the general public, anyone involved would start to understand what climate scientists are concerned about. Perhaps by carefully thinking about the formation of smaller and local trans-disciplinary groups, the discussions of reality in relation to climate change can be freshly approached and extended by networks. As Leopold also stated, 'our land is also part of our ethics and therefore we are individually responsible for it'. If then people feel excluded from the land, that exclusion tends to breed distance and in turn more ignorance.

Perhaps another strategy to help the public understand nature is to explore aspects of the cycles that naturally occur in the atmosphere itself. Although

Ballangee in the Yorkshire Sculpture Park, Eco Action, 2008. Public Field Trip (Photo: Jonty Wilde)

DBF 42, *Elektra Ozomene*. Brandon Ballangee and Stanley Sessions. IRIS print on Watercolor Paper. A cleared and stained specimen of a multilimbed Pacific Tree Frog from Aptos, CA, USA, 2008

many people have a rudimentary idea about evaporation, the molecular realities of climate science are often very unfathomable for them. When Silvia Hostettler was resident artist at the GIG in Lausanne, she became fascinated with the process of photosynthesis. By basing her artwork on the Stomata or structural tissue pores that open and close in the leaf and stem epidermis of plants, the chemical gas exchange that produces our breathable oxygen is explained. As she writes: 'when the red lips of the stoma open, this colour showed that the gene was active so I tried to create something broader and more comprehensive in order to reach the public.' In the final installation, which was shown in the actual foyer of the scientific institute, an enormous human size image of the Stomata covering over 500 petri dishes was used to draw in the viewer to observe the concept of the process itself. In this case the audience included students and staff from the other faculties in the University. Such works cause cross-stimulations and conversations outside of the 'closed doors' of plant genetics. Not that compelling and understandable visualization are not made by the scientists themselves, but they tend to favour the use of front-end traditional illustrations for the public understanding of science. Because our plants are part of us, in this case scale was a very apt strategy, as it tends to place an

extenstionalist view onto the viewer and intensify the role of the witness to the simpler cycles of life. In relation to the scale of such an enormous problem such as climate change, each one of the above artists[13] have chosen to focus on interdependent parts of the problem, which is perhaps the only responsible tactic to take, given that our very idea of nature may need to be re-composed. Meanwhile organizations like the Green Museum.org, know that 'nature', which is constantly in the process of being assembled and reduced through decay, is just waiting to be better understood and reintegrated back into our lives.

Conclusions

Recently, French theorist Bruno Latour[14] commented on the dangers of blundering and rolling into the future of progress, without a glance backwards in order to be critically reflective. He recommends that we drop out of the ideology of progress, which is 'like a state of fumbling in the dark', and instead talk about a new composition for the future. Perhaps modern culture with its divided mainstream and grass roots histories may also need to be rethought in the same way! For years important fabricators and architects claimed that replacement was the answer. Buckminster Fuller for example, insisted that 'you never change things by fighting the existing reality, to change something, build a new model that makes the existing model obsolete.' (Fuller 1981). This may not be the goal of the sustainability conscious art groups who are concerned about ethical and social responsibility of restoring and maintaining nature, rather than producing so many 'things'. Fuller's so called 'design science revolution' encouraged scientists, inventors, architects and designers worldwide to focus their energies on creating and introducing artefacts that would 'enrich human life and bring about a world that works for 100 % of humanity' (1981). However, what would he say now? Artists already worry about the definitions of 'progress' and the consequences of making artefacts out of 'suspect' materials and wasting energy. Surely, the fact that 30 percent of the world now lives off 70 percent of the worlds diminishing resources is the most difficult task at hand. Scientists can already provide any adequate proof of how this fast-forward trajectory of progress will create many more serious problems, but people go astray when claiming that such information dictates what policies ought to be established to deal with the problem. Third-person scientific methods/ perspectives cannot displace insights drawn from first and second person perspectives, an attitude favoured by the scientists working with 'intergral' biology.[15] Therefore, the first person, i.e. the artist's ethical views may also be worth considering.

In this essay I would thus like to conclude with some main questions and attempt to answer them with some 'food for further thought'. These should only be seen as tasty morsels that relate to the examples of the artists and scientists who care and whom I drew with me into this essay. How can an artist become a stronger and more responsible member of these idealistic scientific teams where mobility, energy, and communication systems are needed to reinvent our lives in such overcrowded urban futures? In doing so how can he or she maintain as well as widen his or her provocative place in the world as an artist? How can art and science help each other?

Certainly, as the examples in this essay show, artists can attempt to fill the ubiquitous gap of post-reflection that can attract public support and as well promote more responsible art engagement on an ethical and social level. For this endeavour we may need the following: an improved level of access to the developers of technological progress and their openness to be critically reflective about the definition of progress; further levels of education to discuss how to provide stimulating artworks for the few as well as the many, including those who are perceptually impaired or who 'just cannot listen'; find provocative strategies about the impact of biotechnology, and join forces with the scientists and their facts to tackle the challenging task of climate sustainability. These are only some of the pressing social problems that cannot be resolved by a single disciplinary perspective.

Certainly the attitudes towards the funding of transdisiplinary projects should be reviewed and changed. Also, our educational institutions would have to become places where the relations between science, art, society, and the environment are discussed in open forums and where public discussions about how to solve problems on such large scales are featured. Artists would need to help scientists learn about the semiotics of communication and teach

Hello Weather by Andrea Polli: at the ZHdK Media Campus, Zurich
<http://www.flickr.com/photos/andreapolli/4114176759/
sizes/l/in/set-72157608296740040>

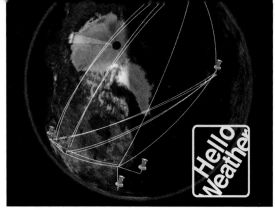

Hello Weather: Diagram of the network showing the location of international stations

them how critical engagement can be made within the sphere of fabrication itself and scientists would have to help artists understand their research and include art/science debates in their specialist's conferences. It would seem that knowledge can no longer be artificially fragmented and a multidisciplinary approach must be encouraged so that better decisions can be made. In science, more efficient use of resources and equipment (like microscopes etc.) is already occurring by sharing them across disciplines, but dynamic changes in knowledge construction are needed. As Basarab Nicolescu, Président, Centre International de Recherches et Etudes Transdisciplinaires (CIRET) 1 posits, 'Transdisciplinary education has its origins in the inexhaustible richness of the scientific spirit, which is based on questioning, as well as on the rejection of all 'a priori' answers and certitudes contradictory to the facts. At the same time, it revitalizes the role of deeply rooted intuition, of imagination, of sensitivity, and of the body in the transmission of knowledge. Only in this way can society of the twenty-first century reconcile effectiveness and affectivity.'[16] Only by learning more about the social impacts of scientific discovery and the ethical discourses surrounding the scientific process can we become privy to the problems of scientific neutrality and the ethical questioning of scientific profit.

There is a living worldwide movement out there and it is slowly growing, but now it is really time for the younger generations' network to step up to the plate, its time for some more grass roots action. This will happen only if artists move beyond the boundaries of the 'me' generation and the post-modern dilemma, into a role where art can again become a larger part of life. Humanity is already a seething interdisciplinary mass that thrives on imagination, and art is the only creative and reflective filtering process that can offer to tackle issues of social and ethical responsibility with paradox, irony, and satire. We all can contribute, even in small interdependent local ways.

Notes

1 In this on-line interview with Okwui Enwezor, ('I have a global antenna' by Rutger Pontzen), Enwezor also stated that access to the camera awarded many artists the chance to have a more socio/political voice. Accessed 20 December 2009, <http://www.vmcaa.nl/vm/magazine/002/artikel004/index.html>.

2 These quotes are extracted from an interview by the author at the ECSITE Conference in 2009. The Leonardo Da Vinci Museum, Milan, Italy. For more information see <http://www.ecsite-conference.net/content/user/File/programma.pdf>.

3 For more information see the article entitled: 'Carrotmob hits Berlin', from 2009, accessed 20 February 2010, <http://www.young-germany.de/news-verwaltung/news-singleview/article/7692e9da2f/carrotmob-hits-berlin.html?no_cache=1>.

4 For a more comprehensive overview about the future of HCI technology from such companies, see the Microsoft report on HCI, accessed June 2009, <http://research.microsoft.com/en-us/um/cambridge/projects/hci2020/biography.html>.

5 The following websites are relevant topics from the artists in this section: Monika Codourey accessed 24 March 2010, <http://www.constanttraveller.info>. Tiffany Holmes, accessed 24 March 2010, <http://tiffanyholmes.com/?page_id=108>. Margaret Tan, accessed 24 March 2010, <http://jy1970.tripod.com/id22.html>.

6 Since 2007 I have been working with the Neurobiology lab at the University of Zurich, accessed February 2010, <http://www.zool.uzh.ch/Research/Neurobiology.html>.

7 As early as 1960, neuroscientist Paul Bach y Rita believed in brain plasticity and sensory substitution and so he was interested to take one sense and use it to detect another: in this case use the sense of touch on the tongue to visualize the surrounding for naïve blind subjects. For more information see: *The Neuroscientist,* accessed 20 January 2010, <http://nro.sagepub.com/cgi/content/abstract/2/5/260/>.

8 The artists and further information about the works in this section can be found at the following websites. Jill Scott, accessed 24 March 2010, <http://www.jillscott.org>. Ellen Levy, accessed 24 March 2010, <http://www.complexityart.com/Reviews/sciart.htm>. Luca Forcucci, accessed 24 March 2010, <http://lucalyptus.com/newSite/sinstall.html>.

9 SymbioticA, located at the University of Western Australia is a centre of excellence in the Biological Arts and one of the most outstanding groups, which I regard as truly Art/Biology transdisciplinary. See <http://www.symbiotica.uwa.edu.au>.

10 The largest share of the GMO crops planted globally are owned by the US firm Monsanto. In 2007, Monsanto's trait technologies were planted on 246 million acres (1,000,000 km2) throughout the world, a growth of 13 percent from 2006, accessed 24 March 2010, <http://www.monsanto.com>.

11 The most comprehensive report on the discoveries of the art/science team, Stanley Sessions and Brandon Ballengee, can be found in the BBC Earth News, accessed: Feb 2010. <http://news.bbc.co.uk/earth/hi/earth_news/newsid_8116000/8116692.stm>.

12 All three artists have related website on-line which feature documents about these attitudes: Hina and Matthias, accessed 24 March 2010, <http://regrowingeden.ch>. Catts and Zurr, accessed 24 March 2010, <http://www.decordova.org/decordova/exhibit> 2003/pigwings.html. Brandon Ballengee / Arts Catalyst project, accessed 24 march 2010, <http://www.artscatalyst.org/experiencelearning/detail/ballengee_ecoventions>.

13 See the site and blogs of these artists for more information about these issues: Amy Lipton, accessed 24 March 2010, <http://ecoartspace.blogspot.com/2009/02/mnncom-interview-with-amy-lipton.html>. Andrea Polli, accessed 24 March 2010, <http://eyebeam.org/hello-weather>. Sylvia Hostettler, accessed 24 March 2010, <http://www.sylviahostettler.ch>.

14 Extracted from a recent interview by the author with Bruno Latour during a conference at the Swiss Museum of Transport, Lucerne, entitled 2010: *The Large, the Small and the Human Mind,* 8th Swiss Biennial on Science, Technics + Aesthetics, accessed 24 March 2010, <http://www.neugalu.ch/e_bienn_2010.html>.

15 Integral ecology is a mixture of methods to study organisms in their environments by taking their complexity into account, acc. 24 March 2010, <http://www.integralecology.org/vision>.

16 More information on transdisciplinary education by Basarab Nicolescu can be found at the following website, accessed 24 March 2010, <http://www.hent.org/transdisciplinary.htm>.

References

Bourdieu, P & Haacke, H 1995, *FREIER AUSTAUSCH: Für die Unabhängigkeit der Phantasie und des Denkens,* S. Fischer Verlag, Berlin.

Blanke, O 2010, in this Volume, pp. 106.

Carr, A & Hancock, P 2003, *Art and aesthetics at work,* Palgrave Macmillan, Basingstoke UK.

Cramerotti, A 2009, *Aesthetic Journalism: How to inform without Informing,* Intellect Books, Bristol UK.

Doidge, N 2007, *The brain that changes Itself: Stories of personal triumph from the frontiers of brain science,* Viking Press, New York.

Dourish, P 2001, *Where the Action Is: The Foundations of Embodied Interaction,* MIT Press, Cambridge/MA.

Edwards, D 2009, *Artscience: Creativity in the post-Google generation,* Harvard University Press, Cambridge/MA.

Ewen, SWB & Pusztai, A 1999, 'Effect of diets containing genetically modified potatoes expressing Galanthus nivalis lectin on rat small intestine', *The Lancet,* vol. 354, no. 9187.

Fuller, B 1981, *Critical Path,* St. Martin's Press, New York.

Greenfield, A 2006, *Everywhere: The dawning age of ubiquitous computing,* Peachpit/New Riders, Berkeley/CA.

Haraway, DJ 1997, *Modest_Witness@Second_Mellennium.Female_Meets_OncoMouse: Feminism and Technoscience,* Routledge, New York.

Hilbeck, A 2010, 'Reconnect', in *Transdiscourse: Mediated Environments,* vol. 1, A Gleiniger, A Hilbeck & J Scott, (eds), Springer, Wien/New York.

Keller, EF 2000, *The century of the gene,* Harvard University Press, Cambridge/MA.

Leopold, A 1949, 'Land Ethic', in *A Sand County Almanac,* Oxford University Press, New York, also downloadable, accessed 24 March 2010, <http://aldoleopold.net/About/LandEthic.pdf>.

Lippard, LR 2008, *October,* MIT Press, winter, no. 123, pp. 105–106, published online 4 February, accessed 24 March 2010, <http://www.mitpressjournals.org/doi/abs/10.1162/octo.2008.123.1.105?journalCode=octo>.

Lipton, A 2002, 'Ecovention', in S Spaid, *Ecovention: Current art to transform ecologies*, Greenmuseum.org, Contemporary Arts Center, Cincinnati/OH, accessed 24 March 2010, <http://green-museum.org/c/ecovention/sect11.html>.

Lucier, A 1965, Music for a Solo Performer, performed at the EMF Institute, accessed Feb 2010, <http://emfinstitute.emf.org/exhibits/luciersolo.html>.

Nicolescu, B 2002, *Manifesto of Transdisciplinarity*, State University of New York (SUNY) Press, New York, transl. from the French by K-C Voss.

Richards, P 1998, 'From London to Naga, Interactive Art at the Exploratorium', in *Art@science*, C Sommerer, & L Mignonneau (eds), Springer, Wien/New York, p. 215.

Wise, SP & Desimone, R 1988, 'Behavioural Neuropsychology: Insight into seeing and grasping', *Science* 4, vol. 242, no. 4879, pp. 736-741.

ART AT THE END OF TUNNEL VISION:
A SYNCRETIC SURMISE Roy Ascott

Syncretism can provide dynamic coherence to competing world-views, scientific paradoxes and emergent cultural practices: The most urgent eco-necessity today is the re-design of ourselves. We are at the final frontier of knowledge: consciousness, a calling for the imaginative deployment of new technologies of communication, computing, as well as chemistry and older somatic practices. Our sense of Being and of Time is changing. The creative navigation of our seamless Variable Reality calls for technological and noetic development of the orthodox sensorium and of the previously excluded second-order senses. The 'iMeme' refers to the multiple self, our generative and distributed personas, emerging in telematic society and in the evolution of consciousness. Concomitant with the views of Bohm, Grobstein, and Blofeld that all matter, animate or otherwise, is mind, this paper concludes that art has a new responsibility towards the creative unfolding of reality.

While historically, paradigmatic shifts in culture have taken centuries to come into being, resisted by unremitting orthodoxy, the fifty years of the post-modernist turn can be seen as relatively brief, largely because much of the previous modernist concern for progress and innovation was retained under its much touted mask of relativistic non-linearity. Whatever its merits as a distinctively new theory of culture, post-modernism's pessimism, negativity and dystopian anxiety made it unsustainable. Essentially life denying and bleakly existential, it added nothing to the spiritual dimension of life, and regarded claims of originality, creativity and authenticity with tired cynicism. While this relativism was temporarily refreshing after the restrictive orthodoxy of Enlightenment

dogma, artistic grand narratives, and scientific fundamentalism, it failed to find a way of bringing dynamic coherence to competing world-views, scientific paradoxes and emergent cultural practices. By contrast, this paper proposes that a wholly syncretic approach to these issues can provide a way out of the postmodern blind alley. Thus, syncretism is the guiding principle of this present text; it brings light to the end of tunnel vision.

In reviewing the way we are now, we see that our planet is telematic, exhibiting dense and inclusive global connectivity; our media is moist, exploiting the convergence of digital and biological systems; our mind is technoetic, with technology extending our cognitive repertoire, and in some cases transforming consciousness; our sensorium is extended by prostheses that are bringing about a new faculty (that we have called cyberception) (Ascott 1994); our individual identity is becoming multiple with the creation of avatars and alternate personas; our body is transformable both in physical and virtual terms; our reality is variable, seamlessly connecting an evolutionary environment of manifold worlds; our substrate in the construction of our reality is at the nano level, interfacing the material and immaterial conditions of being. In consequence, art will become progressively more syncretic, or risk losing entirely its already enfeebled social and spiritual significance.

This paper proposes the following perspective: That which calls for reconstruction, realignment, and regeneration in the world, is the re-designing of ourselves as the most urgent eco-necessity. To undertake this task, we must recognise what it is like to be

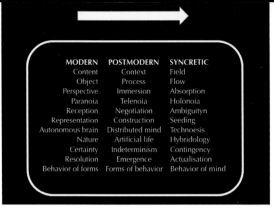

The cultural shift from modernism to syncretism in the arts, Roy Ascott, 2008

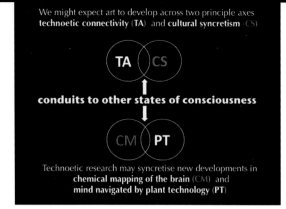

Conduits to consciousness with the affordances of brain chemistry and plant technology, Roy Ascott, 2008

living in a variable reality, where chance and change in both the environment and in ourselves is constant, and where the future is perceived as unpredictable, unreliable, and uncertain. We are constantly updating, remodelling and re-inventing ourselves, seeking new relationships, new realities, and in effect new orders of time and space. We have an open-ended culture, which is evolving and transforming at a fast rate, and an art that develops strategies of ambiguity, contingency, and play. The outcome of all this is that the binary opposition between real and virtual realities no longer holds, and that what could be called the 'Grand Western Illusion' of the individual human brain creating its own isolated consciousness is exploded in favour of a recognition of a connected intelligence seeking fuller access to the primordial field of universal consciousness.

We had previously adapted our sensibilities to fit into what were considered to be separate boxes, containing the real, the virtual, and the spiritual, each serving a separate ontology. But now we can see the emergence of a syncretic coherence, accelerated as much by the revival of ancient somatic practices, as by the imaginative deployment of new technologies of communication, computing, and chemistry. This has led us to see the variability of presence, consisting in physical presence in ecospace, numinous presence in spiritual space, telepresence in cyberspace, and not least vibrational presence in nanospace.

This has brought us to what might be considered as the final frontier of knowledge: consciousness. This paper proceeds from the view that consciousness may be a non-material, irreducible given, that mind may not be an epiphenomenon of the brain, and that evolution may have produced the brain as an organ of access to consciousness, whose domain is infinite and timeless. From this it is argued that the evolution of technoetic systems, whether somatic, nano, digital or pharmaceutical, will extend this capacity.

Whether or not the reader accepts these precepts, it must generally be agreed that 'Being' is not what it used to be. The issue of a complex variable reality compounds the uncertainty of being-in-the-world, since we recognise that all states are transient and all boundaries permeable. On reflection we shall no doubt celebrate our own de-grounding, finding that to be globally distributed and connected gives amplitude to our hypercortex and to mind at large. We can be seen as engaged in a kind of telematic teleology, where the end justifies the media. The media being moist: chemical as much as digital, a kind of accelerated alchemy. As a consequence of the technologies of the mind and of the body, our sense of 'Being' and of 'Time' is changing. Through our engagement with Second Life and other virtual worlds and virtual communities, we have a sense not simply of being distributed asynchronically but of being multiple, with multiple identities, effectively rejecting the existential single-self. Increasingly we recognise that we inhabit phase-space, and live in non-linear time. From this can be extracted new thinking about the nature of design, which may apply at all levels and in all aspects of living. We argue that the best design is that which infuses intention with the improbable, and results in the unforeseen. Design should be transformative of both object and user. Transformative Design is the design of flows. This is to speak of 'field-design'. Dealing with the world, and ourselves, in terms of field-design will generate entirely new forms of behaviour and communication, new systems and structures. The field-effect will be felt in architecture, entertainment, learning, post-industrial production, prosumer resources, geophysical boundaries, and psychophysical features. Field-consciousness will generate a more collaborative, syncretic approach to solving our problems and designing new initiatives. Science, art and technology must rethink their interactions: Field-design will be the integrative outcome.

Thus, syncretic design may become a methodological imperative. Syncretism may serve us in understanding the multi-layered worldviews, both material and metaphysical, that are emerging from our engagement with pervasive computational technologies and post-biological systems. The application of syncretic thinking has distinct and positive effects. It accelerates technoetic evolution, destabilises orthodoxies of thought, challenges representation, fights dogma, confronts materialism, demands participation, hybridizes identity, smoothes social interaction, and re-orders time and space. In response to developments in new media art around the turn of the millennium, the term 'moistmedia' was coined to signal the emergent confluence in media art of (wet) biological processes and (dry) computational systems (Ascott 2000, pp. 44–49). Ten years later, the term is insufficient unless it is understood to include bio/neuro/geo/chemico/cogno/nano/astro/pharmo and psycho media. Moistmedia that transits the spectrum of wet and dry, natural and artificial, embodied and distributed, tangible and ephemeral, visible and occult.

The outcomes that can be expected of strategic field-design will be moist and immaterial, grounded and numinous, embodied and distributed. They will lead to the cultural coherence of intensive interconnectivity, quantum coherence at the base of our world-building, and spiritual coherence of our multi-layered consciousness. In processes concerned with designing for techno-creativity, five principal pathways are proposed: amplifying thought (concept development); designing identity (self-creation); seeding structures (self-organising systems); making metaphors (knowledge navigation), and sharing consciousness (collaborative processes).

As we noted earlier, the most urgent need in our present eco-economy is the design of the self, the reconstruction of self and identity. Central to the process is the extension of the senses. This applies to both technological and noetic development – eventually to be syncretised as technoetic research.

Technological Development

In the case of technological development of the sensorium, much research over recent decades has taken place in the extensions and prostheses. Of the many sensory and cognitive fields to which this applies, examples are found in the areas of direct brain–computer communication, silent speech interfaces, self-perception, and body-swapping systems.

Brain computer interfaces can augment sensory and cognitive functions in the field of neuroprosthetics with applications aimed at restoring damaged sight, hearing, and mobility. Due to the cortical plasticity of the brain, signals from implanted prostheses can, after adaptation, be handled by the brain-like natural sensor or effector channels. Research is also producing new personal interfaces for human computer interaction. You think a command, and the object of your attention behaves accordingly. A neurochip for rat brains is being developed that can act as a prototype for the development of prostheses in the human brain's hippocampus, whose function it is to encode experiences for storage as long-term memories elsewhere in the brain. Thomas DeMarse has used a culture of 25,000 neurons taken from a rat's brain to fly a F-22 fighter jet aircraft simulator. The cultured neurons began rapidly to reconnect themselves to form a living neural network. The cells were arranged over a grid of 60 electrodes and used to control the functions of the simulator. This 'brain in a dish' experiment as this was popularly tagged, may lead to 'living computers, that may someday be used

to fly small unmanned airplanes or handle tasks that are dangerous for humans, such as search-and-rescue missions or bomb damage assessments' (Wireheading 2004). However, it is perhaps in the gaming industry that the most spectacular popular developments in telekinesis are to be found, with headsets that allow the player's thoughts to control events in an onscreen scenario.

Another sensorial development to be noted is the development of technology that facilitates communicating with silent speech. For example, research at NASA by Dr. Charles Jorgensen is producing a Silent Speech Interface (NASA 2008) that 'enables speech communication to take place without the need to emit an audible acoustic signal'. Using sensor data from such parts of the human speech system as neural pathways, or the brain itself, the SSI produces a digital representation of speech that can be routed into a communications network (Universal Communication 2008). Similar research is taking place in many other parts of the world. (Silent Speech Communication 2010).

The manipulation of the senses in terms of self-perception, and the experience of the illusion of body-swapping, is conducted in research at the Karolinska Institutet in Stockholm. Here cognitive neuroscietists, led by Henrik Ehrsson, have succeeded in making subjects perceive the body of others as their own (Petkova and Ehrsson 2008).

A more dramatic example of technologically-based challenges to our understanding of cognition and the senses is found in neurocardiology. In western science the heart has long been understood to be the seat of emotion and the brain the organ of thought. The research of J. Andrew Armour challenges that. He argues that the heart's brain is an intricate network of several types of neurons, neurotransmitters, proteins and support cells like those found in the brain proper. Its elaborate circuitry enables it to act independently of the cranial brain – to learn, remember, and even feel and sense. Armour proposes that the heart communicates with the brain and body in four ways: neurologically through transmissions of nerve impulses, biochemically through hormones and neurotransmitters, biophysically through pressure waves, and energetically through electromagnetic field interactions. In short, Armour's assertion is that the heart may use these methods to send the brain extensive emotional and intuitive signals, and may be the intelligent force behind intuitive thoughts and feelings (Armour and Ardell 1994).

Noetic Development: The recognition and exercise of second-order senses

Aristotle identified just five senses: sight, hearing, touch, smell, and taste. Neuroscience provides a further six: pain, balance, proprioception, kinaesthesia, sense of time, and sense of temperature. These we may call senses of the first order.

But in the case of what we would call 'Second-Order Senses', seen as unorthodox and marginal by established, mainline science, technoetic systems (digital, somatic, pharmaceutical) are designed to enable us to traverse further reaches of consciousness, to access psychic states, and extend our spiritual awareness. An approach to these senses may be likened to second-order cybernetics that mirrors this field phenomenon with its co-dependence of the observer and the observed. Spiritual states and psychic awareness require first-person participation, second-order senses, and the emergent faculty cyberception. But it may be asked: How are second-order senses exercised and perhaps developed further? The process involves following what might be called the Tao of second-order senses, the stages of which appear to be: stepping outside the Enlightenment box; quiet contemplation of the mind; examining doctrines dubbed esoteric, hermetic, occult; recognising one's self as a field phenomenon; extreme curiosity about spiritual aspects of non Western cultures; reaching for the metaphysical implications of biophysics; seeking shamanic contact; imbibing traditional psychoactive substances such as ayahuasca within a ritual framework; imaging intensively the development of second-order sensibility; and designing with divination rather than with deliberation.

As a consequence of technoetic evolution we are rebuilding the self, and as such, we are no longer a single-self organism. We are each engaged in constructing and syncretising many selves, many personas and mixed identities. Progressively we shall become more permeable and transparent, both at the edge of the mind and in the materiality of the body. This will apply not just to others, but also to ourselves and our own self-realization. The deeper we go into ourselves, the more selves we discover. We recognise that the self is not fixed but generative. We are in an endless state of becoming, in an endlessly variable reality.

Second-order senses (a Western taxonomy), Roy Ascott, 2008

Syncretic reality chart, Roy Ascott, 2007

Richard Dawkin's *Meme* comprises a unit of cultural ideas, symbols or practices, which are transmitted from one mind to another. The term *iMeme* is introduced in this paper to refer to the multiple self, generative and distributed, emerging in telematic society and in the evolution of consciousness. In the Net, the *iMeme* carries an element of the identity of its originator, self-replicated and bifurcated into a variety of new forms and behaviours, and transmitted through telematic networks and in new media spaces. Just as with the computer, avatars will only become intelligent when they can feel and their feelings are transmitted throughout the distributed multiple identity of a user. But this raises the question: is Second Life a subsystem of a singular reality system or is it a second-order system, from which subjects of the first order reality system can observe their originating selves?

As Paul Grobstein has argued in *Constructing Reality and the Self* (2002):

'The nervous system is an exploratory device, continually building and revising models of the world. The experience of "seeing" involves two distinguishable (and dissociable) stages. The first consists of the detection and analysis of visual input; the second involves the conscious "experience" of seeing. This bipartite arrangement means that much of what an individual's nervous system does is not experienced by that individual, raising unsettling questions about the meaning of "self", and of "reality". What one experiences is not with any certainty what is "out there", but is instead an interpretation of sensory inputs done by a set of processes within the nervous system which are always to some degree unpredictable and unknown. The "self" is a complex and changing blend of things, some of which at any given time one internally experiences oneself, and at other times does not.'

Some (particularly Eastern) philosophies argue that we think ourselves into being. We can say further that the technoetic evolution of consciousness consists in our desire to think ourselves into being multiple, and to exercise a distributed sense of presence. John Blofeld asserts that all matter is mind. 'I am at one with my Taoist and Buddhist friends in believing that, at a more nearly ultimate level of experience, it is found that all matter, animate or otherwise, is mind. It is because we cling blindly to the seeming facts reported by our senses – despite all the denials of their ultimate validity voiced by scientists and mystics alike – that we fail to perceive, much less accept, this truth' (Blofeld 1959).

In the philosophy of David Bohm, thought is participatory. It produces and shapes our perception of reality:

'We see reality according to our thought. Thought is constantly participating both in giving shape and form and figuration to ourselves and to the whole of reality. Thought doesn't know this. Thought is thinking that it isn't doing anything. The difficulty is to see that thought is part of this reality. We are not merely thinking about reality – we are thinking it' (Bohm 1996).

Just as *Participation Mystique* (Levy Bruhl) consists in the subject being unable to clearly distinguish himself from the object but bound to it by a direct relationship which amounts to partial identity, a relic of the original non-differentiation of subject and object, the primordial unconscious state, a kind of integral empathy, prior to representation. So we might usefully identify *Participation Cybernetique* as consisting in a telematic immersion in the Net, which leads to multiple identity, syncretic representation, and absorption in the global consciousness.

The solid objective world that we see is a representation of the particles that we recognise as constituting quantum phenomena at another level of resolution. But we can reasonably ask whether these sub atomic particles are a representation of something beyond that which is *really* real? If the three-dimensional, solid world of ours depends on our consciousness, what existed in the world before human consciousness had evolved?

These questions, seen often as ethereal or unconnected to the gross realities of daily toil and conflict that seem to mark out our days in what has been called *The Decade from Hell* (Serwer 2009), are none the less pertinent and perhaps fundamental to our negotiations with the nano field, and our constructions of reality in the quantum domain. We seem to be building a syncretic reality, which calls for a syncretic art, the emergence of which will depend on a new understanding of being-in-the-world, and a new responsibility towards that creative unfolding.

References

Armour, JA & Ardell, L (eds), 1994, *Neurocardiology,* Oxford University Press, New York.

Ascott, R 1994, 'The Architecture of Cyberception', first publ. in *ISEA '94 – The 5th International Symposium on Electronic Art,* Helsinki, accessed 3 October 2009, <http://caad.arch.ethz.ch/teaching/praxis/ss99/sources/ascott.html>.

Ascott, R 2000, 'The Moist Manifesto', in Konrad, H & Kriesche, R (eds), *Kunst – Wissenschaft – Kommunikation,* Springer, Vienna/New York, pp. 44–49.

Blofeld, J 1959, *The Wheel of Life: the autobiography of a Western Buddhist,* Ryder, London.

Bohm, D 1996, *On Creativity,* Nichol, L (ed.), Routledge, London.

Ehrson, H 2008, 'Scientists produce illusion of body-swapping', *Neurosciences,* Karolinska Institutet, accessed 3 October 2009, <http://ki.se/ki/jsp/polopoly.jsp?l=en&d=21944&a=75292>.

Grobstein, P 2002, *The Brain's Images: Co-Constructing Reality and Self,* Serendip, Bryn Mawr College, accessed 3 October, 2009, <http://serendip.brynmawr.edu/bb/reflections/upa/UPApaper.html>.

NASA 2008, *Ames Technology Capabilities and Facilities,* accessed 3 October, 2009, <http://www.nasa.gov/centers/ames/research/technology-onepagers/human_senses.html>.

Petkova, V & Ehrsson, H 2008, 'If I were you: perceptual illusion of body swapping', PLoS ONE, accessed 2 January, 2010, <http://www.plosone.org/article/info:doi%2F10.1371%2Fjournal.pone.0003832>.

Serwer, A 2009, *Time Magazine,* vol.174, no. 22, pp. 22–28.

Silent Speech Communication, vol. 52., no. 4, 2010, pp. 270–287, accessed 23 February 2010, <http://www.sciencedirect.com/science/journal/01676393>.

Universal Communication 2008, Keihanna Research Laboratories, Kyoto. accessed 3 October 2009, <http://klab.nict.go.jp/document/khnlab_en.pdf>.

Wireheading 2004, *'Brain' in a dish acts as autopilot, Living Computer,* University of Florida, accessed 3 October 2009, <http://wireheading.com/misc/artificial-brain.html>.

ART AND SCIENCE RESEARCH TEAMS? SOME ARGUMENTS IN FAVOUR OF A CULTURE OF DISSENT Nina Zschocke

artists and art historians in labs

While working on a project concerned with theoretic models describing interrelations between higher cognitive and perceptual processes, I was kindly welcomed as a research fellow at laboratories of neu- robiology – as an 'art historian in lab' one might say. While I was officially treated as a 'visiting scholar', it was quite clear that I was seen rather as a more or less exotic 'guest' than as a 'collaborator' in a strict sense. A different perspective onto a shared field of interest, for example disparate ideas about what 'art' might be and which kind of questions were interest- ing to ask when dealing with works of art, were some of the issues that lead to a difference in research objectives and hypotheses. And even though I had entered a fruitful cross-disciplinary exchange of ideas on certain problems, I discovered that I was simulta- neously acting as a kind of ethnologist (trying out the limits of participant observation) (Geertz 1987). My attention was drawn to certain aspects of labo- ratory practice that did not appear to be of central concern to my hosts. For example, at a certain point, I was much more interested in the variables of a par- ticular experimental set up – instructions given to the probands – than in the actual empirical results. The observations I made, however, helped me to redefine my hypothesis, to refocus the project, to search more efficiently for literature, to make new interesting contacts and so on. Furthermore, during my stay, I profited from discussions gaining new insights into the methodological background of certain argu- ments, helping me to overcome some of the problems that I had experienced when struggling with scientific papers on my own. I would argue therefore, that for me, as an art historian with a specific, well defined research project, a temporary research residency at a scientific laboratory was a highly valuable experience which complemented, but could never substitute, the independent research outside of the lab, 'in the library', in a department of art history, but also in the exchange with other scholars from various disciplines.

Academic institutions offer research residencies quite frequently to researchers from other, yet related, backgrounds within the sciences. Compared to being employed as a regular team member, it is one notable characteristic of such programs that they facilitate relatively short-termed encounters. Due to the tem- porary nature of the involvement (and with an eye to career management) the invitee is in many cases best advised to design his or her own project in a way that sets it apart from the main corpus of work done at the host institution, while still profiting from the interaction. This draws attention to the 'guest' as an individual agent, a migrant between spaces and subcultures of knowledge production, striving for originality in his or her work. Driven by the requests of an individual project that transgresses institutional boundaries, the research fellow engages in network- ing as an activity, tying together several diverse lines of research. On the other hand, this can lead to a potential conflict between the host and invitee concerning the character of the work to be done and the goals to be accomplished during the residency. Of course, there are cases without such conflicts, and scientific labs inviting artists for short-term residen- cies often show an exceptional generosity in offering resources. While looking at encounters between indi- viduals engaged in rather distant fields of research,

Figure 1: Takis in his studio at MIT's Center for Advanced Visual Studies, 1968. Credit: Davis, Douglas: Art and the Future. A history/prophecy of the collaboration between science, technology and art. New York 1973, p. 127

Figure 2: Robert Irwin and James Turrell in UCLA's Anechoic Chamber, 1969. Credit: Davis, Douglas: Art and the Future. A history/prophecy of the collaboration between science, technology and art. New York 1973, p. 166

this paper will not attempt to make a general comparison between 'artistic' and 'scientific' practices or products as such. And I will not analyse strategies to effectively overcome differences and controversies. Rather, I will argue in favour of a 'clash of interests' and advocate a 'culture of dissent'. A conflict of interests and opinions not only allows the guest's project to express a critique of the host's assumptions and practices, but also increases the chance for the interaction to produce unpredictable results, and therefore, something 'new'. Referring to some cases of collaboration between artists and scientists, the paper will examine different modes of encounter. Furthermore it will discuss 'errors' and shifts of meaning occurring in the processes of reproduction, translation and negotiation as a potential for creative processes.

acculturation and difference

Vassilakis Takis, who worked at the Center for Visual Studies at MIT as a visiting researcher/artist (1968–69), makes a rather schematic distinction between his collaborations with 'technicians' (or 'technical helpers') and 'scientists' in an interview with Douglas Davis, yet some of his arguments are still worth considering. (Fig. 1) Takis argues that 'technicians' have only little political influence, as they serve the

ruling establishment by producing new technologies conceptually designed or commissioned by a coalition of government and science (or by the industry). As the government is fundamentally dependent on the loyalty of science, it is threatened by a potential collaboration between scientists and artists: 'I feel that governments now depend on scientists. I cannot imagine any government, therefore, happy about the scientist-artist union. On the other hand, technicians represent no threat. If we work with the technicians alone, we will only publicize the products of government' (Takis in: Davis 1973, p. 130). The artist, stereotyped by Takis as driven by a 'bohemian, agitated' mind (and therefore as fundamentally critical of the government) is for that reason 'bribed into harmless cooperations'. In fact, long-term employments of artists and scientists (or engineers) as regular team members and as permanent collaborations carry the risk today as much as in the 1960s of transferring onto the artist (and the scientist) a lab's dependency on – for example private – funding and its obligatory loyalty to a sponsor or commissioner. Even in cases where the artist is not morally or politically corrupted and instrumentalized to 'white-wash a company's tainted image', (Shanken 2006, p. 11). There is a natural tendency – caused by the logic of commissioned work and industrial funding – that the artist is employed as an innovator guaranteeing better designs or innovative solutions that are in line with the economic and political goals of the commissioner. Apart from a direct involvement of artists in the development of commercial products, or in serving industrial demands, it can be added that also artists who 'harmlessly', that is for non-political, playful, aesthetic projects, reuse new technologies that originally had, for example, a military purpose, are successfully deployed to distract public attention from the fact that a technological research program is, for example, funded by the ministry of defence. All this constitutes considerable problems for large parts of the genre of 'new media art'. Ars electronica's *futurelab* explains on its website that it 'brings together the two concepts [of the artist's atelier and the researcher's lab] in a single workspace [...] in which the tone is set by activities of transdisciplinary teams and which, depending *on the demands of a particular assignment,* is continually being reconfigured as a lab-atelier or atelier-lab.' It is added that a recent 'shift in emphasis from the creation of interactive installations for the Ars Electronica Center [that is: art related work] to carrying out projects for and in collaboration with commissioning clients clearly attests to an opening of the market for this kind of know-how.' The website further promotes the lab's projects in the field of 'media art and architecture' as examples that 'can be used as a visual *expression of a company's corporate culture* while fulfilling certain functional tasks' (Ars Electronica Futurelab 2009).

Even though it doesn't seem appropriate to morally stigmatise such media labs and their members for having found and developed economically success-

ful niches, which in the best case have offered quite satisfying jobs, it remains notable that a long-term integration of artists into design teams carries the risk that the artist becomes a media designer in a process of acculturation. This process is accelerated by the fact that media art is often taught at schools of applied arts and sciences that – with a number of exceptions – tend to train their students mainly to become successful participants in a design economy. Correspondingly, quite a number of projects labelled as 'media art' are more interesting if regarded as, for example, pure apolitical 'interface design' (Zschocke 2007). However, only when art, in contrast to commercial product design, resists an all too easy and undisturbed consumption, and instead acts as a resistor, undermining established concepts or everyday modes of behavior or perception, then art can even hope to 'change the direction that science and technology [and at least in that sense: society] will take' (Malina 2006, p. 18). Caroline Jones, for example, sees that one role of art to create new uses and extensions of technology should be to '*make it strange* again, in the best aesthetic fashion, helping us to take the measure of our techno-bodies and their sensorium' (Jones 2006, p. 43). Accordingly, it can be argued that, 'cultural hacking' (or 'détournement') is not only a strategy of critique and subversion, but also a chance for the emergence of something new.

In this context, it might be helpful to replace Takis' stereotyped 'technicians' and 'scientists' with a distinction between 'basic' and 'applied' research. When James Turrell and Robert Irwin participated in Los Angeles County Museum's *Art and Technology Program,* namely in a collaboration with the experimental psychologist Edward Wortz in 1968/69, the scientist's research was funded by NASA because of its potential value for future applications. Still, the psychologist's work at that time can be described as 'basic research', as Wortz's following statement shows: 'The problem as stated in science has no more of a specific "why" than anything else.' When undertaking an experiment, he said that he was 'just curious about what happens' and commented on the collaborative work he did with Irwin and Turrell: 'The line on which we were able to relate with respect to utility was pretty good. This doesn't mean that these things don't have any value, eventually, but that they were produced for themselves. And the utility is either indeterminable or irrelevant' (Wortz in: Davis 1973, p. 164). Correspondingly, basic or 'pure' research can be defined as being driven by a curiosity or interest in a scientific question. The main motivation is to expand human knowledge, not to create or invent something. There is no direct and obvious commercial value to the discoveries that result from basic research (even though often it is essential to name potential future applications in order to successfully apply for funding). Applied research, in contrast, is designed to solve practical problems rather than to acquire knowledge for knowledge's sake. One might say that the goal of the applied scientist is to improve the human condition (for example the health of the population) or to serve certain – often either military or economic – goals of a commissioner. Artists and scientists engaged in artists-in-labs programs might in fact differ most in their long-term concerns and objectives. While the scientist involved in basic research will most likely describe the redefinition of theories or the productions of new epistemological

objects as his/her primary aim, the visiting artist might consider such new insights gained at the lab mainly as a *tool* or a stepping stone towards the realization of his/her own project.

James Turrell and Robert Irwin shared with Edward Wortz's *basic research* a general interest in certain characteristics of perception and mental states (compare Wortz in: Davis 1973, p. 164). The experiments conducted by the three collaborators using an anechoic chamber had the goal of producing a subjective experience that is comparable to internal states achieved by certain practices of meditation (Fig. 2). Wortz stresses, that 'the quality of the internal state that we wanted to achieve was the real objective'. And he adds: 'In one sense, our concern about the internal state has spilled over into another endeavor, which both Bob and Jim are pursuing with great enthusiasm, and that's an assault on our environment in general' (Wortz in: Davis 1973, p. 165). In other words, Wortz's research that was paid by Nasa and Garret/Airesearch to – in the long run – produce insights into the requirements for 'creating a safe, functional, and productive environment' to 'enable humans to safely and effectively live and work in space' (Rando et al. 2004, pp. 5–9; NASA habitability 2009) provided input for artistic projects that aimed at quite the opposite: pseudo meditative experiences or internal states that are (like meditation) contrary to work efficiency. Turrell, for example, in many of his installations, confronts observers with a visual paradox. They perceive light as a seemingly solid material, with oscillations between two- and three-dimensionality, undermining the efficiency and reliability of everyday vision. Turrell writes: 'In my work, you become aware that the act of observing can create color and space' (Turrell 1998, p. 180; Zschocke 2006, pp. 143–173).

While the artistic 'goals' of James Turrell and Robert Irwin differ fundamentally from the aims of research on space habitability and have no direct value to space travel, their work could be partly integrated into basic research done by Edward Wortz. It might even be argued that while collaborating with the two artists, the scientist himself was to a certain extent 'distracted' from projects more adequate for serving NASA and 'seduced' into experiments in line with his own interest in more general problems concerning the nature of consciousness (and in buddhist practices of meditation in particular – Wortz co-founded a Buddhist center in 1969). It remains open for discussion however, whether one wants to see the artists' projects as harmless byproducts and 'decorations' of scientific experiments conducted in the framework of cold war space research – or as an exploration of artistic media and modes of experience that led to independent articulations in line with a changing public awareness at the end of the 1960s. While Turrell's installations, for example, can not be described as 'political' in any strict sense, they engage visitors in experiences and trigger reflections that lead away from the need to function properly in a commercial world, in military action, or on a mission to expand America's extraterrestrial territory.

cleaner fish, virus and cuckoo

There are many advantages to looking at 'artists' and 'scientists' as individual agents and not as members of different 'species' (even though Darwin's decon-struction of the classic notion of species sheds light on such a comparison, by describing it as subject to constant mutation and change). Nevertheless, when searching biology for pointed metaphors (or carica-tures), the concept of 'symbiosis' comes into view. First of all, different types of symbiosis vary between sporadic or permanent relationships. Furthermore, distinctions are made between 'mutualism' (each partner derives a benefit) and 'parasitism' (the para-site benefits while the host suffers damage). Ecology might give both partners of artists-in-labs programs hope by telling us that a symbiosis is most likely to be mutual, that is profitable for both, when host and guest are maximally distinct from each other. However, it is questionable and to be decided case-by-case whether artists and scientists really do have widely differing living – or rather, working – require-ments. On the other hand, the notion of the 'parasite' could be metaphorically applied to artists joining labs in order to use new technologies (and technological support) otherwise not available for their purpose, namely in projects that are not profitable for – or even harm – the host institution. This might be a model for artistic strategies of subversion. Parasites are closely linked – by the metaphor of the virus – to the concept of the 'hacker' – like the 'pirate', a romantic (self-)image of the activist. When a 'virus' is transmitted, the host becomes infected and, in some way or other changed. Brood parasitism, as practiced by the cuckoo, who places her eggs in another bird's nest, is another metaphor. Finally, ecology labels the class of those relationships between two organisms where one benefits but the other is unaffected as 'commensalism'. However, the commensal relation is often observed between a larger host and a smaller commensal. Furthermore, while the host organism is unmodified, quite frequently the commensal species – the *cleaning fish* is a classic example – shows great structural adaptation, in other words: the guest adapts to the host. Seen in this light, one could argue, that as long as artists at a media or technology lab (take MIT as an example) produce aesthetic, yet harmless applications of new technologies, that were (like many VR applications in the 1980s and 90s and nanotechnology today) developed for military purposes, they popularize the government's products. The beautiful cleaner-fish serves as a caricature of the artist who serves, decorates and follows the host (a predator) where it goes, profiting from his/her own loyalty and willingness to adapt. In any case, it is clear that subversive practices can only be realized by artists who remain independent in the long-run. Therefore, long-term integration of artists into design teams has to be set apart from short-term 'migration' and visits, as enabled by 'artists-in-labs' programs that are at best funded by public or art institutions. Again, I use the notion of the 'guest' in order to do justice to the independence of the visitor of a lab (the symbiosis is facultative, the relationship is beneficial but not essential for the invitee's economic survival) and to the temporary – or sporadic – nature of the interaction.

networking and assemblies

In an essay examining the role of new concepts of space, time, and matter in art, Jesús R. Soto adresses the danger for an artist when he becomes 'absorbed' in scientific research, as, 'indubitably, the world of scientific knowledge, which is as fathomless as the world of art, *may absorb him* in its complexity and prevent him from returning to his art' (Soto 1994, p. 228). I have myself observed cases in which humanists 'have become' neurobiologists. In accordance with Soto one could argue that such a transformation (or 'absorption') has not been unfavorable for the subjects themselves but rather for their activity in their original field of study (Soto 1994, p. 228). In the worst case, we are left not with a cross-disciplinary exchange of thought and practice but with only a quantitative change, for example with more neurobiologists and less humanists – or with more scientists or engineers and less artists. In contrast, the concept of 'networking' as an activity could be applied to any research that does not follow a beaten path. Rather than resembling a linear, target-oriented search, it is characterized by a strolling movement, roaming freely amongst different spaces of knowledge production while developing a project. These activities might include research residencies, the consultation of scientific publications, day visits to labs and other institutions, experiments and conversations with scientists – all driven by the current specific interest of the artist / researcher. Hans-Jörg Rheinberger has pointed out that an experimental system is designed to let something 'new' occur, yet the scientist does not know exactly what this newly emerging 'thing' will be. On the other hand, Rheinberger argues, it is indispensable to have a 'vague idea of what one is looking for in order to be surprised' (Rheinberger 2007, pp. 87–89). The same might be true for the artist's movement through the fields of science while looking for, and gathering, concepts, models, questions, materials and methods. He or she moves about in a landscape like a hunter-gatherer, evaluating its resource structure and interpreting and following tracks. Here, 'networking' appears as a metaphor for an activity of exploring and experimentally linking a growing number of sources, contact points, and encounters with human and nonhuman agents.

Shifting the focus again from the activity of an individual to the events that lead to new scientific 'articulations', Bruno Latour challenges us to address scientific – and in particular laboratory – practice as 'an assembly, a gathering, a meeting, a council'. Pointing out that 'any epistemology is political', Latour describes scientific laboratories (and networks of connected laboratories) as types of assemblies, which gather a 'public' around 'things' or disputed 'matters of concern' (Latour 2004, p. 221; Latour 2005, p. 23). He refers to a description of science as 'restricted and circumscribed to tiny, fragile, and costly networks of practices'. Bruno Latour argues that science study has reimplanted objectivity 'into plausible *ecosystems*' and that 'the truth conditions epistemologists had looked for in vain inside logic had finally been situated in highly specific truth factories' (Latour 2009, pp. 1–2). What Latour calls 'Dingpolitik' requires 'objects' or 'matters of facts' to give way 'to their complicated entanglements and to become matters of concern' while 'opinions' turn into 'issues' (Latour 2005, pp. 23, 41). When examining the complex set of technologies, interfaces, platforms, networks and mediations allowing for 'things' to be made public, the two central questions to be asked are, Latour insists, 'Who is to be concerned' and 'What is to be considered?' (Latour 2005, p. 16). Latour discusses the requirements for scientific results or 'articulations' to be 'interesting' (and not redundant) in another argument, where he contrasts

Karl Popper's principle of falsification with the one defined by Vinciane Despret and Isabelle Stengers. He points out that in contrast to Popper's theory, the 'real risk to be run is to have the questions [the scientist was] raising *requalified* by the entities put to the test.' Therefore, what is to be falsified is 'not just the empirical instance of the theory, but also the theory, the very research programme of the imaginative scientist, the technical apparatus, the protocol.' In other words, as the Stengers-Despret criterion requires the experimental scientist to constantly check whether he is asking the right questions or whether he has to change his laboratory settings, it requires the scientist 'to jeopardize his privilege of being in command' (Latour 2004, p. 216). Following this argument, we can observe that not only the 'entities put to the test' but also the artist as a 'guest' in the laboratory, having a different interest or concern, might – as a (participant) observer – ask unexpected questions and point to aspects of the experimental set-up that are so much part of the discipline's tradition that they escape the scientist's critical eye. The exotic visitor imports into the lab conflicting interests and, generally, dissent. Seen in this light, 'mutualism' (a benefit for both, host and guest) might therefore not be positively correlated with a similarity in interest and research objectives or with a full acculturation of the guest, a merging of interests – but, to the contrary, with an increase in distance between the two partners's points of view. Once we look at the meeting of artists and researchers in a scientific lab as assemblies of individual agents with different interests engaged in a trial, disputing 'matters of concern' – or as a multi-disciplinary assembly around a 'thing' or 'problem', the relevance of the preservation of different interests and points of view becomes apparent, both as regards to politics (democracy) and creativity (understood as the emergence of something unexpected).

translations, copies and mistakes

As Roger Malina points out, 'ideas flow in a number of ways between intellectual communities' (Malina 2006, p. 16) – or, rather, they are transferred and translated by individuals that themselves migrate between different subcultures and spaces. 'Concepts' can be considered as key to intersubjective understanding as everyone is supposed to be able to 'take them up and use them' (Bal 2002, p. 22). As abstract representations of an object, concepts are thought to be tools of intersubjectivity and to facilitate discussion on the basis of a common language. However, as Mieke Bal points out, like all representations, concepts are 'neither simple nor adequate in themselves. They distort, unfix, and inflect the object.' To take up concepts and use them is, as Bal argues, 'not as easy as it sounds, because concepts are flexible: each is part of a framework, a systematic set of distinctions […].' Rather than being objective descriptions or labels for an object, concepts involve interpretative choices. Bal: 'In fact, concepts are, or rather *do*, much more. If well thought through, they offer miniature theories […].' And as such, when used in an analysis, 'concepts can become a third partner' in the interaction between researcher and his object (Bal 2002, pp. 22–23). Furthermore, whenever concepts are transferred (or 'travel') between individual scholars or between disciplines – between 'science' and 'art' for example – 'their meaning, reach, and operational

value differ' (Bal 2002, p. 24). Theories, models, and concepts function differently in different settings and when integrated into different traditions. Accordingly, the recontextualization of whatever has been adopted from science has to be well understood as an activity by the artist. A scientific concept, theory, or model is reframed in an event that has an effect on the concept itself and on its possible uses (Bal 2002, pp. 134–135). When examining individual artistic projects and their relation to a concurrent or previous contact with science, it is interesting to ask not only 'Who is concerned?', and 'What is to be considered?' but also more specifically; 1. What was it exactly that the artist picked up from 'the lab', 2. 'How was it applied to and how did it interact with the artist's own matter of concern', 3. How did the result – a work of art for example – relate to the original 'interests', 'things', and 'concepts' encountered at the lab? Regarding the use of scientific concepts or theories in art, Mieke Bal's arguments help us to focus on artistic practices of selection, reproduction, translation, framing, and materialization, changing the function of gathered theories and tools, producing distortions and thereby new meaning.

While the observation that processes of copying and translation produce errors, and thereby changes and unpredicted results are relevant for any transfer between disciplines, it plays a particular role in the work of the German artist Carsten Nicolai. The scientific models and theories themselves that are collected by Nicolai during his strolls through different spaces of knowledge production, center around various aspects of indeterminacy and self-organisation evolving from certain characteristics of reproduction,

2002, p. 63). While in many of Nicolai's projects such as *bausatz noto ∞* (1998), 'indeterminacy' is introduced by uncontrollable variables of interaction, such as the behaviour of visitors, projects like *XERROX* make use of indeterminacy as an effect of self reproduction.

The copying machine (referred to by a word combination of the brand name 'Xerox' and 'error') functions as a metaphor. More closely, the compositional concept refers to characteristics of digital reproduction. The basic observation behind the XERROX project is that a multiplication of the process of copying – to make a copy of a copy of a copy – and the conversion from one digital format to another produces interpolations, small changes in the encoded sounds and thereby indeterminacy. Fragmentations and errors, 'glitches' that characterize low resolution, mask the relation to the original. For the realization of the project, an application for a sequencer was programmed that produces (simulates) these effects of copying. While in many pieces the sound material used by Nicolai / alva noto consists of pure sinusoidal tones produced by oszillators, on *XERROX VOL. 1* (2007), he works with sounds recorded in commercial and public spaces. The 'sample transformer' is used to manipulate the recordings that have been dissected into audio fragments, de-familiarizing and transforming them into something that manifests its connection to the original everyday sounds only suggestively (Pesch 1998, p. 328). As the record's accompanying text points out, 'simplifications and deformations lead to a gradual loss of the copy's relationship to the original and can result in a substantial change of meaning. Although elements of the source remain,

the original message dissolves amongst the white noise of reproduction' (Nicolai 2007). The process of copying itself becomes a creative tool.

XERROX. vol. 1 is not self-referential in the (modernist) sense that it would primarily reveal characteristics of the computer program as a medium. Neither do the tracks 'represent' a scientific theory in such a way that it could be retranslated or even used to calculate. It is also not 'a machine that plays itself', nor can the compositions included in the album be reduced to or sufficiently 'explained' by those principles of information processing that played a part in its production. Instead, a concept, gathered from a scientific paper, feeds into a compositional strategy. However, the theory itself has been translated – and thereby transformed. Here, a theoretic model 'gathered' from a scientific source is translated into an artistic concept, into a compositional strategy and then into a computer program – an application for a sequencer (which again serves to translate digital data into sound).

The concept of 'active mutation' (the emergence of new patterns due to processes of reproduction and translation) was not only applied to manipulate digital sound, but also describes what happened to the scientific concepts themselves when selected and integrated into the artist's work. Scientific models were changed (distorted) in the course of a number of translations: into different languages, – for example from math to words, by an artist who debates with musicians, into strategies, into technical applications, and into media and materials. Step by step in the course of new experiments, some elements of

Figure 6: alva noto and Ryuichi Sakamoto with ensemble modern: utp_, 2008, performance. Credit: alva noto and Ryuichi Sakamoto with ensemble modern: utp_, 2008, dvd (raster-noton), video still

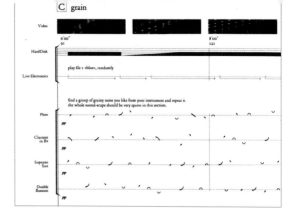

Figure 7: alva noto and Ryuichi Sakamoto with ensemble modern: utp_, 2008, score, detail. Credit: alva noto and Ryuichi Sakamoto with ensemble modern: utp_, 2008, dvd (raster-noton), accompanying booklet

the original theory might be lost or hidden behind other layers. This might be particularly true for collaborative projects like the audio-visual composition and performance *UTP_* (2008) commissioned by the city of Mannheim. Together with Ryuichi Sakamoto and the *Ensemble Modern,* Carsten Nicolai engaged in a many-voiced negotiation of musical and visual composition. An assembly, gathered not to produce 'epistemological objects' but to produce an experience (Rheinberger 2001). (Fig. 6) In a first phase of the project sound 'particles' produced by the members of the ensemble and their instruments were recorded. These were then applied to a 'model', based on the rasterized structure of the 17th century city map of Mannheim, but also to the classical music tradition of the city (Mannheim school), and to serial music (Nicolai 2009; Pesch 2009, p. 88; Pesch 1998, p. 327; Stockhausen 1963, p. 452). The composition was transcribed (translated) into a musical score, which became interpreted again and functioned as the basis for improvisation (Fig. 7).

Projects that don't shy away from dissent but rather assemble groups of actors with different interests, and that involve many facets of translating between languages of different disciplines and traditions, lead to unexpected experimental results. 'Errors are

fundamentally human', Nicolai says, 'this sets us apart from machines' (Wahjudi 2001, p. 224). And he comments somewhere else: 'In research, mistakes and disruptions lead to the most interesting results. It is nothing else that happens in art' (Nicolaus 1999, p. 83). But just as the 'previously unknown' does not show itself in science to the unskilled experimenter, so the 'new' presence in the final composition or performance does not result from any kind of sloppiness. Carsten Nicolai's work, in contrast, aims at a maximum of precision when weaving a web to catch something unexpected (compare Rheinberger 2007, p. 85).

artists-in-labs – are we invited to join?

It is interesting to observe that 'individuals "enter" the art/science/technology networks from different nodes, or roles, in various stages in their own work or lifetimes.' And that 'often, it is the creative friction between dissimilar disciplines that create the conditions for unanticipated outcomes' (Malina 2006, p.16). What I want to point out, however, is that these 'outcomes' or 'products' still trigger different responses when regarded from the perspective of one protagonist to the next. A project might seem more valuable to the 'guest' than to the 'host', more interesting when seen as 'art' than as a contribution to 'science', or something similar. This directs our attention towards the potential audience – or to the problem of choosing or creating the context of the work's presentation. Even though the boundaries between disciplines are subject to slight, but constant changes, the question for which public an artist (or art historian) in lab produces and exhibits (or publishes) has not become obsolete. What language is chosen, which expectations and interests are addressed, which knowledge is referred to? To which field of research should the project contribute, in which framework should it be interpreted? While artists, driven by their interest, act as independent networkers transcending disciplinary boundaries, their works might call for an equally multi-disciplinary audience. However, when Carsten Nicolai states that it occurred to him relatively late that everything he had done would be regarded and commented within different 'branches', it indicates that disciplinary boundaries, however translucent and perforated they might be, continue to have a persistent effect on defining 'interest groups' (Pesch 1998, p. 326). Contemporary art seems to have crossed all boundaries and broken all rules that might have existed. But it appears to me that when it comes to its relationship to science, there is a hesitation to confront visitors with scientific material so sophisticated that it does not lend itself to an immediate understanding. However, it might be worthwhile to remain an 'exotic guest', not only in the lab, but also in the museum. Since it doesn't have the function to simply popularize science, I believe art should not worry about whether it is asking too much of its audience. It does not have to be 'easy' and interaction does not always have to be 'playful' and 'intuitive'. Rather it is challenging to be offered

links or traces that can be followed. Notes made by an artist in the margins of a scientific paper can function as cross-disciplinary links or 'tunnels' to other fields and practices of research. It can act as an invitation to the audience to get involved, to not only 'play' but add its own articulations. While one strategy after an experiment might be to return to a classic concept of an exhibit that stands on its own, hiding the 'frictions' and disputes that were part of its production under a smooth surface, another option could be to allow the audience to open the 'black box' of a work, revealing some aspects of the 'experiments' and 'assemblies' that brought it to life. As much as individual scientific 'results' can only be artificially and painfully isolated from research as an ongoing process that involves multiple agents and variables, so also works of art, when re-implanted 'into plausible ecosystems', can only gain when references are given to the 'networks of connected laboratories' and to the 'tiny, fragile, and costly networks of practices' that were essential to its emergence (Latour 2009; Latour 2005).

References

Ars Electronica futurelab 2009, accessed 9 March 2010, <http://www.aec.at/futurelab_about_en.php>.

Ars Electronica human nature 2009, accessed 9 March 2010, <http://www.aec.at/humannature>.

Bal, M 2002, *Travelling concepts in the humanities: A rough guide,* University of Toronto Press, Toronto.

Davis, D 1973, *Art and the Future: A history-prophecy of the collaboration between science, technology and art,* Praeger, New York.

Geertz, C 1987, ''Aus der Perspektive des Eingeborenen': Zum Problem ethnologischen Verstehens' (1977), in C Geertz, *Dichte Beschreibung: Beiträge zum Verstehen kultureller Systeme,* Suhrkamp, Frankfurt/M, pp. 289–309.

Haglund, M 2005, 'The air between the planets: on the audio and visual works of Carsten Nicolai', in *carsten nicolai: anti reflex* (exhib. cat.), Schirn Kunsthalle, Frankfurt/M, pp. 24–38.

Ikegami, T & Hashimoto, T 1995, 'Active mutation in self-reproducing networks of machines and tapes', *artificial life,* vol. 2, no. 3, pp. 305–318.

Jones, C 2006, 'The mediated sensorium', in C Jones (ed.), *Sensorium: Embodied experience, technology, and the contemporary art,* MIT Press and MIT List Visual Arts Center, Cambridge/MA, London, pp. 5–49.

Latour, B 2004, 'How to talk about the body? The normative dimension of science studies', *Body & Society,* vol. 10, no. 2–3, pp. 205–229.

Latour, B 2005, 'From Realpolitik to Dingpolitik or how to make things public', in B Latour & P Weibel, *Making things public: the atmospheres of democracy,* MIT Press, Cambridge/MA, pp. 14–41.

Latour, B 2005, 'Spinoza lecture II: the aesthetics of matters of concern', in B Latour, *What is the style of matters of concern?* Van Gorcum, Amsterdam, pp. 26–50, accessed 9 March 2010, <http://www.bruno-latour.fr/articles/article/97-STYLE-MATTERS-CONCERN.pdf>.

Latour, B 2009, 'Spheres and networks: two ways to reinterpret globalization', a lecture at Harvard Graduated School of Design, Feb. 17, 2009, in *Harvard Design Magazine,* no. 30, pp. 1–2, accessed 9 March 2010, <http://www.bruno-latour.fr/articles/article/115-SPACE-HARVARD-09.pdf>.

Malina, R 2006, 'Welcoming uncertainty: the strong case for coupling the contemporary arts to science and technology', in J Scott (ed.), *artists-in-labs: Processes of inquiry,* Springer, Vienna/New York, pp. 15–23.

Nakaya, U 1954, *Snow crystals: natural and artificial,* Harvard University Press, Cambridge/MA.

NASA habitability and environmental factors division homepage 2009, accessed 9 March 2010, <http://hefd.jsc.nasa.gov>.

Nicolaus, F 1999, 'Carsten Nicolai: Ich bin ein Mauerspringer', art, no. 12, pp. 78–83.

Nicolai, C 2007, *alva noto: XERROX vol. 1* (record envelope), raster-noton, Chemnitz.

Nicolai, C 2009, 'Two and a Half Questions with Carsten Nicolai', *Headphone Commute,* accessed 9 March 2010, <http://reviews.headphonecommute.com/2009/09/08/two-and-a-half-questions-with-carsten-nicolai>.

Obrist, HU 2002, 'Hans Ulrich Obrist im Gespräch mit Carsten Nicolai',

in raster-noton (ed), Carsten Nicolai: auto pilot, Berlin, pp. 59–68.

Pesch, M 1998, 'Mehr als Crossover – Techno, House und neue elektronische Musik in der aktuellen Kunst', U Groos & M Müller (eds) *Jahresring* (Make it funky. Crossover zwischen Musik, Pop, Avantgarde und Kunst), no. 45, pp. 325–329.

Pesch, M 2009, 'Transfer and transformation: strategies in the oeuvre of carsten nicolai', *parachute* (électrosons/electrosounds), vol. 107, pp. 81–93 (and other essays in the same volume).

Rando, CM, Baggerman, SD, Duvall, LE & Martin, 2004, 'Habitability in Space', *Human Factors and Ergonomics Society Annual Meeting Proceedings, Aerospace Systems,* San Diego, CA, pp. 5–9.

Rheinberger, H-J 2001, *Experimentalsysteme und epistemisch Dinge,* Wallstein-Verlag, Göttingen.

Rheinberger, H-J 2007, 'Über die Kunst, das Unbekannte zu erforschen' (2006), in P Friese, G Boulboullé & S Witzgall (eds), *Say it isn't so: art trains its sights on natural sciences,* Kehrer Verlag, Heidelberg, pp. 83–91.

Schröder, B 2007, 'Immateriality and the ephemeral in the works of Carsten Nicolai', in *carsten nicolai: static fades* (exhib. cat.), Haus Konstruktiv, Zurich, pp.143–145.

Shanken, EA 2006, 'Artists in industry and the academy: collaborative research, interdisciplinary scholarship, and the interpretation of hybrid forms', in J Scott (ed.), *artists-in-labs: Processes of inquiry,* Springer, Vienna/New York, (pp. 8-14).

Soto, JR 1994, 'The rolsdddsf – fdsf e of scientific concepts in art', *Leonardo: Journal of the International Society for the Arts, Sciences and Technology,* vol. 27, no. 3, pp. 227–230.

Stockhausen, K 1963, 'Texte zur elektronischen und instrumentalen Musik', Bd. 1, *Aufsätze 1952–1962: zur Theorie des Komponierens,* 'Es geht aufwärts', vol. 9, D Schnebel (ed.), DuMont Schauberg, Köln, pp. 391–512.

Turrell, J 1998, 'Crater Spaces', in *James Turrell: The other horizon,* (exhib. cat.) MAK, Wien, pp. 156–186.

Wahjudi, C 2001, 'Carsten Nicolai: Systeme aus Klang und Bild', *Kunstforum International,* Bd. 155 (Der gerissene Faden: Nichtlineare Techniken in der Kunst), pp. 222–224.

Zschocke, N 2006, *Der irritierte Blick: Kunstrezeption und Aufmerksamkeit,* Wilhelm Fink, München.

Zschocke, N 2007, 'Kollaborieren und Plappern: Das Internet als Testfeld relationaler Ästhetik', *31: Das Magazin des Instituts für Theorie der Gestaltung und Kunst* (Paradoxien der Partizipation), December, no.10/11, pp. 77–84.

CASE STUDIES: ARTISTS-IN-LABS 2007–2009

THINK ART – ACT SCIENCE: MEETING ON AN EQUAL LEVEL Irène Hediger

'One has to do something new in order to see something new'
(Lichtenberg 1998 (1789–1793), p. 321).

Introduction and History

The artists-in-labs project was established in 2003 at the Zurich University of the Arts (ZHdK) in Switzerland. The main aim of the project was to explore the interface between art and science in the lab context and conduct research about the innovative potential of such collaborations.[1] In 2006 the project turned into a program with the goal of providing a long-term development of alternative ways for artists to learn and be inspired by their interactions with scientific research and to respond with contemporary art practices. In cooperation with the Swiss Federal Office of Culture OFC, the project *Sitemapping*[2] now offers four Swiss-based artists nine-month residencies every year. Our aim as facilitators is to create permeable and mutable collaborations between artists, scientists and engineers. We believe that new fields of creative research and knowledge can be produced when artists and scientists meet and this encounter can 'open up' other forms of knowledge. By sharing the specific expertise of artists with those of scientists, these nine-month encounters provide a solid ground for productive interrelations to develop. We focus on the process of the experience, rather than on the production of a finished artwork. With this platform we give artists and scientists the opportunity to share their preliminary forces, and to explore their creativity and desire to understand and represent unknown outcomes.

The ideology behind the artists-in-labs program is based on the provision of many different immersive experiences inside the various cultures of scientific research. This requires that the artists have actual 'hands on' access inside the lab itself, as well as attend relevant lectures and conferences concerning topics in physics, engineering, computer and life sciences. Artists are inspired to develop their content and their interpretations accordingly, and to reflect upon the impact of technical and social issues of scientific inquiry on the general public. The program also helps scientists to learn about current methodological, aesthetic and communication developments in the arts and to gain some insight into the world of contemporary art and media. The potentials of this encounter with 'the other' often inspires both the artists and the scientists involved to reflect upon their role in society, their knowledge production and methodology, as well as the creative levels of their research.[3]

As a program, we are able to provide access to science labs across Switzerland that might otherwise be out of reach for artists. Over the past few years we have worked with seventeen Swiss science institutes constituting a large selection of scientific disciplines, spanning the life sciences, natural sciences, physics, computer and engineering sciences. Our goal is to maintain long-term collaborations with our hosting institutions and gradually add new partnerships with science institutes, thereby expanding the scope of our research into geographical and cultural situatedness. A good residency has to be carefully facilitated and each side has to be very clear about their expectations. To provide such productive settings for the collaborations requires preliminary negotiations and contracts. We expect the science institutes to pro-

Midway-Meeting with Irène Hediger and directors of the science labs for 2009: Marco Conedera, Martin Pohl, Luca Gambardella and researcher Oliver Kannape

vide the artist with a working space and access to the relevant scientific information and infrastructure, similar to any other scientific team member. A main pre-requisite is that scientists in the hosting institution must provide some hours of tuition for the artist, and, depending upon the institute's possibilities, the artist should be encouraged to attend relevant lectures and conferences. We have developed a residency framework with milestones, deliverables and workshops providing time and space for reflection for all involved – the artists, the scientists and the facilitators. With the flow of recommendations and feedback that comes from the artists and scientists, the program is constantly improved. Besides setting up the collaborations with various science institutes that we have already worked with in the past, we visit new laboratories and negotiate with the directors of these institutes in order to extend the scope of science institutes geographically and thematically.

The Selection Process

Once a year, at the beginning of June, we publish a call for project proposals on our website <http://www.artistsinlabs.ch>, in major Swiss newspapers, national and international online-platforms and send it to our database contacts. The artists then submit project proposals for one of the science institutes and research focus of their choice. In 2006 we received 31 applications and by 2009 the response to our call rose to over fifty project proposals. Currently, the selection process is multi-staged. In the first stage the Swiss artists-in-labs co-directors together with the lab directors and others in the research team, review and select a short-list to be later presented to

a jury. We select up to five potential finalists from the applications based on the following criteria:
· Well-researched concepts, which consider the current state of research in the chosen science environment as inspiration for artistic production.
· The level of originality, innovation and interpretation in the proposal.
· The project plan for the residency including prototype production.
· The ability to communicate ideas, processes and methodologies.

In a second stage, the co-directors conduct a face-to-face interview with each of the short-listed artists together with the scientists. The third stage of the selection, conducted by a jury of artists and curators is based on the following criteria:
· The relevance of the content in the artist's proposal in relation to general development in the arts.
· An overview of the artist's past work and an assessment if such a residency would be a good development for the career of the artist.
· The potential to distribute or expand the project in the future or to exhibit or publish the result.
· Relational issues such as scope of art disciplines covered by the award of the stipend, geographic location problems, commitment and gender.

Arts) held in 2008 in Singapore. Thanks to the organiser's permanent efforts to expand and develop the artists-in-labs programme, the programme has become a valuable enrichment for the Swiss cultural landscape.

Notes
1 See 'Strategy of the Federal Council for an Information Society in Switzerland', February 18 2002, p. 1, as retrieved from <http://www.bakom.admin.ch/themen/infosociety/00695/index.html?lang=de>.
2 'Strategy of the Federal Council for an Information Society in Switzerland', February 18 2002, p. 4.

SCIENCE LABORATORIES

Life Sciences:

INSTITUTE OF INTEGRATIVE BIOLOGY (IBZ) | ETH ZURICH

CENTER FOR INTEGRATIVE GENOMICS (CIG) | UNIVERSITY OF LAUSANNE

WSL SWISS FEDERAL INSTITUTE FOR FOREST, SNOW AND LANDSCAPE RESEARCH, BELLINZONA

EAWAG: THE SWISS FEDERAL INSTITUTE OF AQUATIC SCIENCE AND TECHNOLOGY, DÜBENDORF

Cognition & Physics:

THE BRAIN MIND INSTITUTE (BMI) | EPFL, LAUSANNE

THE HUMAN COMPUTER INTERACTION LAB (HCI LAB) | INSTITUTE OF PSYCHOLOGY
UNIVERSITY OF BASEL

THE PHYSICS DEPARTMENT AT THE UNIVERSITY OF GENEVA | CERN

PAUL SCHERRER INSTITUTE (PSI), VILLIGEN

Computing & Engineering:

CSEM SWISS CENTER FOR ELECTRONICS AND MICROTECHNOLOGY, ALPNACH

THE NATIVE SYSTEMS GROUP | COMPUTER SYSTEMS INSTITUTE | ETH ZURICH

ISTITUTO DALLE MOLLE DI STUDI SULL'INTELLIGENZA ARTIFICIALE (IDSIA), MANNO-LUGANO

ARTIFICIAL INTELLIGENCE LABORATORY | UNIVERSITY OF ZURICH

INSTITUTE OF INTEGRATIVE BIOLOGY (IBZ)
ETH ZURICH ZURICH

Focus

The Institute for Integrative Biology at the ETH belongs to the Department of Environmental Sciences and consists of ten professorships and two special units. We undertake an integrative approach to investigating biological systems in order to better understand their properties and function. These include native, invasive and novel organisms such as genetically modified organisms (GMOs). Our mission is to increase the understanding of ecological and evolutionary processes affecting interactions among organisms and biological systems; to transmit the knowledge gained so that it contributes to human welfare and the sustainability of natural resources and to educate our undergraduate and graduate students in the relevant fields of research.

Angelika Hilbeck
Senior Researcher

Angelika Hilbeck and her group conduct research on potential environmental impacts of GMOs. She teaches environmental risk assessment in practical courses and has concluded a 6-year GMO Environmental Risk Assessment capacity building project in three countries of the so called 'developing world' (Brazil, Kenya and Vietnam). She is chair person of the European Network of Scientists for Social and Environmental Responsibility (ENSSER).

artists-in-labs: Hina Strüver and Matthias Wüthrich

Hina and Matthias spent 9 months with us in Zurich, Brazil and Vietnam, analysing the situation and making very impressive public performances and installations about GMOs. We were pleased to have artists who could both comment on and interpret the discussions about GMOs in the developed and developing countries like Vietnam or Brazil, (which is on the verge of becoming a developed country). Both artists participated in our teaching and research activities. Their project at ETH was a series of conceptual performances in the large inner patio of the building, which houses the Environmental Sciences department. In these performances they re-enacted the process of transformation of a GMO and the subsequent cell division. This was the theme of a series of two more performances that symbolized the step-by-step development process of the GM plant and lead to a large installation filling the entire glass-roofed patio. The result even included the space above the glass ceiling, a total height of about 10 stories of the entire building complex. Furthermore, the artists travelled to Brazil and Vietnam and spent time there to explore the state-of-the-art discussion about GMOs in these countries and carried out more public performances there.

Our collaboration with Hina and Matthias was true to our original aim in that they indeed commented and interpreted the science around the themes of risk and safety in gene technology. They learnt more through interacting with the scientists from all involved disciplines. In this way, the residency was a very valuable collaboration, and we, the scientists, learnt a great deal about the creative process of conceiving, developing and carrying out a public performance and installation. We were quite impressed with the amount of work needed to construct the installations, the physical effort and use of materials. At

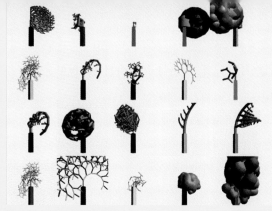

Dr. Angelika Hilbeck, Senior Researcher of the Institute of Integrative Biology at a meeting with Dr. Jill Scott/ZHdK

Virtual garden generated through an interactive platform on genetic engineering: <http://regrowingeden.ch> (Strüver/Wüthrich, 2007)

the beginning and the end of the project, the artists organized social events where many scientists from the entire department were invited and they participated in good numbers. The installation was left in the patio for several weeks and it continued to stimulate discussions among the students and scientists in the building during the duration of the installation. This fact indicates that an artistic interpretation of the GMO controversy meets a great public and social demand. After two articles in the news media reported the project and accompanying events, others were pleased that the artists were at our institute. They enriched our perspective and discussions and helped the public to critically reflect on the controversy surrounding the issue of gene technology.

The artists were most intrigued by the ballistic process of transformation deploying what is called a 'gene gun'. This method and terminology triggered a number of associations and images that the artists translated into an impressive, physically demanding performance: Climbing and swinging in professional climber's gear shooting 'gold particles' coated with foreign DNA symbolized by yellow ribbon balls into the cell symbolized by the huge multi-story, glass-covered patio of the building housing the Environmental Department of the ETH. The diversity of

opinions expressed by the scientists of the institute about this controversial technology then translated into a virtual garden complementing this performance that could be followed by everybody. They used our connections in Brazil and Vietnam to develop similar kinds of art projects in these countries.

Recommendations

We would take another artists-in-labs recipient again any time. We would suggest to focus again on a particular scientific project, and then look for an artist who might fit the project, as this was a very satisfying experience.

www.env.ethz.ch

CENTER FOR INTEGRATIVE GENOMICS (CIG)
UNIVERSITY OF LAUSANNE

Focus
The main goals of our Institute are

1. The development of a first rate research program in the biological sciences.
2. The development of an outstanding teaching program.
3. The development of shared research technologies available to the local research community and beyond.
4. Outreach to main public schools and open door events.

Nouria Hernandez
Director of CIG
Christian Fankhauser
Associate Professor
Laure Allenbach
Lab technician

In Christian Fankhauser's lab <http://www.unil.ch/cig/page8391_en.html> we are interested in the role of the environment on plant growth and development. More specifically we focus on plant responses to changes in their light conditions (e.g. shading by other plants). We study the plant photoreceptors and the mechanisms by which a signal transduction cascade is initiated by their light activation. Light perception leads to specific developmental responses. This is important for the plant to optimise its growth and its reproduction in reaction to environmental parameters. We perform our work with a small plant called Arabidopsis that is particularly well suited for the molecular genetic approach that we are using. We use sophisticated microscopes and LED-incubators for our biological studies and the whole palette of molecular biology and biochemistry tools.

artist-in-lab
Sylvia Hostettler developed a project that really fits the scientific universe in which we are working. She eventually built a 'black box' in which the visitor can enter and where diverse objects were exposed. The main source of light was a window made of recycled Petri dishes, painted on the back to represent a giant stomata. Dispersed in the room, were shiny plastic objects representing undifferentiated plant tissues. The outside of the box was used to expose pictures that were inspired by Sylvia's work with the microscope and by her observations of galls.

Her overall project took into account different scientific topics, which are each embedded in one another. It covered the gene expression field by using the visual of a specialised program called Genevestigator and by creating quite astonishing homemade microarrays. The black box symbolizes the exchange between the outside and the inside (both literally and figuratively) by referring to a special plant structure: the stomata, involved in the gas exchanges during photosynthesis. The plastic shapes inside the box focus on the growth and development of plant tissues and their possible mutations in reference to calli (sort of vegetal tumours). The installation also enhances the importance of light for plants survival by playing with the different light sources. The black box was moreover a wink to our dark room where we perform all our experiments under controlled light conditions. Sylvia's own experimental manipulations are also displayed. She worked extensively

Sylvia Hostettler's art world (Photo: Laure Allenbach, 2008)

Lab world (Photo: CAOS, 2008)

with microscopy, taking images of small collages she made with parts of the plant we use to work with and parts of herself. Sylvia created her Petri window by collecting the used experimental dishes, washing them and painting them. She also tried different materials to research the undifferentiated calli and she worked in the microscope facility. Her preliminary project was presented to the scientists working in the CIG as an informal display and also on a poster during our retreat in Saas-Fee. Moreover people were freely invited to visit her in her art lab as often as they wanted.

The scientists were pleased about this new 'colleague' with whom they could experience a different universe. They were very curious about what would emerge from the interaction between Art and Science and were absolutely enthusiastic about her project. Her office was a breath of fresh air, especially when people wished to quit their benches, pipettes, computers and publications.

The artistic approach of Sylvia had some parallels with the scientific approach including hardship and perseverance. She was not stopped by the difficulties and if the idea was good she just went for it, no matter how long and repetitive it was to achieve it.

Another analogy could be that an idea brings another and so the project moves on step by step. Also she works by trials and improvements as we do.

Recommendations

The time of the residence was long enough for Sylvia to develop her project, but accomplishment takes a lot more time. From this point of view, she didn't have time to finish her project during her residence, but the public presentation of her artwork actually took place in March 2009. We were able to collect enough funds for its achievement, but could more grants be available for some expensive exhibitions that could not be covered by the host institute?

www.unil.ch

99

WSL SWISS FEDERAL INSTITUTE FOR FOREST, SNOW AND LANDSCAPE RESEARCH BELLINZONA

Focus

The Swiss Federal Institute for Forest Snow and Landscape Research WSL focuses on the use and protection of landscapes and habitats with the goal to find the best ways of making use of landscape and forest resources in a responsible way. Furthermore it develops an integrated approach to handling the natural hazards that commonly occur in mountainous countries. Being a part of the ETH domain, the particular function of the research institute is to act as a bridge between pure theoretical science and the practical implementation of scientific findings. A particular strength of WSL consists in providing an interdisciplinary and transdisciplinary research team that is problem oriented and keeps a view to practical solutions.

The Insubric ecosystems group of the WSL is an interdisciplinary research group dealing with different aspects of wildfire history, ecology and management in Switzerland. Main skills of the group are forestry, vegetation science, bio-indication, eco- and palaeo-history, community ecology, etc. and scientific approaches such as experimental field studies; process modelling and risk assessment. The main trigger for hosting an artist of the artists-in-labs program is our awareness about the multitude of possible approaches that exist in viewing, commenting and disseminating science and research results. This is particularly true if science concerns itself with concrete aspects of daily life such as landscape, natural hazards or nature in urban space.

Marco Conedera
Head of the Insubric Ecosystem Research Group

There was a general interest in having an artist in our group because in the different research approaches and methodologies we use observations, controlled experiments and case studies, and we thought it might be interesting to see another way for data to be collected rather than systematic samplings, field surveys, analysis of existing maps or aerial photographs, etc. Our researchers are interested in biodiversity assessment, the modelling of natural processes, visualization of data through thematic maps and reconstructing past land use.

artist-in-lab: Claudia Tolusso

We interacted with Claudia for nine months, and this included teaching, discussing and implementing project activities. At the beginning (first three months) she systematically participated in our internal and external activities in order to have an overview about the daily life here and was surprised about the amount of office work we do in our research group. A basic point of discussion was the scientific terminology and the definition of terms such as model, statistical significance, control and hypothesis in the scientific community. During this phase we gained a lot from the interactions in terms of discussion about our research approach, the way we set up the research strategy and detect specific field of activity and research topics. In a second phase, she worked more independently in order to develop her own ideas on possible artistic performances in connection with our work. Unfortunately, most of the excellent ideas she invented were not finalized, but they do survive in written form in her personal diary! In the last two-months we collaborated to finalize

The WSL Lab in Bellinzona, Switzerland

A slice through time: A giant chestnut containing 400 tree rings that contain the life history of this tree

a proposal and application for funding for three installation/performance ideas and submitted them to the Cultural Department of the city of Bellinzona. These were about seed explosions, green highways and data clouds to be held in three different locations in the city. Called *Vivere i Sensi* (or *Catch the Emotions*) they were accepted for further development for 2010.

Having an artist did not deeply affect our community but sometimes the perspectives of the artists were brought into the discussion and this caused unexpected and basic questions about our terminology or methodological approaches. These had to be answered before the discussion could be continued. What artists and scientists seem to have in common is that initial trigger of a good performance, which in both cases is constituted by a basic level of intuition. In the development phase science has to meet with international established methodological standards whereas the approach to making art seems to be less coded and may therefore develop more freely.

Recommendations

The time will never be long enough. Nine months are too short. Perhaps in the future instead of asking the artist to develop his or her own project, a pool of

scientists and artists could be selected on the base of their curriculum and theme and then asked to submit a common proposal. This could develop during the artists-in-labs time.

www.wsl.ch/bellinzona

EAWAG: THE SWISS FEDERAL INSTITUTE OF AQUATIC SCIENCE AND TECHNOLOGY DÜBENDORF

Focus

Clean water is not self-evident. Switzerland spends billions of Swiss Francs annually to attain this objective. Numerous developing and newly industrialized countries suffer from acute water shortages. Eawag's task as the national research center for water pollution control is to ensure that concepts and technologies pertaining to the use of natural waters are continuously improved and that ecological, economical and social water interests are brought into line.

Multidisciplinary teams of specialists in the fields of Environmental Engineering, Natural and Social Sciences jointly develop solutions to environmental problems. The acquired knowledge and know-how is transmitted nationally and internationally by publications, lectures, teaching, and consulting to the private and public sector. 400 employees are active at the locations in Dübendorf (near Zurich) and Kastanienbaum (near Lucerne). Eawag was founded in 1936 as an information centre for wastewater treatment of the ETH Zurich. It is a Swiss Federal Research Institute which is part of the ETH-Domain. Practically all research projects at Eawag are interdisciplinary or even transdisciplinary oriented. Therefore, exchange not only occurs among biologists, engineers and social scientists, but intensive contacts are also maintained with specialists and decision-makers of the private and public sector as well as professional associations.

Water is at the focus of all research as it is the primary source of life and key to development and prosperity. Its availability as a resource is, however, limited worldwide and qualitatively poor in many areas.

Without sustainable solutions, the struggle for water also increasingly threatens peaceful coexistence. The main focuses of Eawag's aquatic research can be summarized as 'water as habitat and resource' (Aquatic Ecosystems), 'water in urban areas' (Urban Water Management) and 'pollutants in the water' (Chemicals and Effects).

Christopher T. Robinson
Aquatic Ecology

Chris Robinson is an aquatic ecologist specialising in the ecology of alpine and temporary streams, population genetics of alpine insects, disturbance ecology, colonization dynamics, nutrient dynamics, fire ecology, organic matter processing, patch dynamics, and algal ecology. His research program in the Swiss National Park is on:

1. the effects of experimental flooding on streams downstream of reservoirs,
2. the ecological functional assessment of springs, and
3. the long-term biomonitoring of the Macun Lake Biosphere.

Renata Behra
Environmental Toxicology

Renata Behra is an environmental toxicologist using algae as model systems of study. Algae play significant roles in aquatic ecology. Besides being main producers of oxygen, which becomes available to humans and other organisms for respiration, they also provide the food base for most aquatic food chains. Hence, her research is concerned with the protection

Eawag Forum Chriesbach, a sustainable and aesthetic building (Architect: Bob Gysin + Partner BGP, 2006)

Illona Szivak, Environmental Toxicology. These experimental channels show the effects of risk factors i.e. contaminants, ultraviolet radiation on algal biofilms (Photo: Ruedi Keller)

of algae from the impact of environmental pollutants and physical stressors and with the development of tools to assess effects of stressors to algae. Emphasis is put on understanding how effects occurring at various levels of biological organization relate to each other. To that aim, basic and applied studies are carried out combining field studies with microcosm and laboratory experiments. Stressors considered include metals, synthetic nanoparticles and UV radiation.

Chris Zurbrügg
Water and Sanitation in Developing Countries

Chris Zurbrügg is head of Sandec. Sandec is the Department of Water and Sanitation in Developing Countries at the Eawag. It aims to develop, provide and facilitate the implementation of new concepts and technologies in water supply and environmental sanitation, increase research capacity and professional expertise in low and middle-income countries in the field of water supply and environmental sanitation, and raise awareness and enhance professional expertise in high-income countries for water supply and environmental sanitation issues in low and middle-income countries. Sandec's activities centre on problems of sustainable development in economically less-developed countries like Senegal. Its mandate

is to assist in developing appropriate and sustainable water and sanitation concepts and technologies adapted to the different physical and socioeconomic conditions prevailing in developing countries.

artist-in-lab
Ping developed 6 different projects. Two were sculptures based on her perceptions of humans and our very special no-mix toilets. Eawag has developed a no-mix toilet and Ping was rather fascinated about the potential of such a device for human health and environmental sustainability. One project developed from learning the microscope and about microorganisms, macro-organisms and, in particular, the eggs from those organisms. Here she developed a sculpture about her perception of invertebrate eggs, attaching the system to an air pump to vitalize the sculpture.

Ping developed three videos based on her activities in each lab; all document how she perceived the science. One toilet sculpture has been on display in the terrace area of the Eawag and the other two sculptures were completed. Ping had a formal showing of her results to the staff and public of Eawag on 1 December 2008 and the media were also invited to this event. Ping was also interviewed by various art and media interests during the 9 month stay. She was highly interactive with the researchers and their students during her residency. We believe all persons that interacted with Ping during this period gained an acute awareness about how the public is likely to perceive their science: an overall positive experience! Ping was a charming person that interacted well with most everybody she met, and she is a highly determined person that took full advantage of this opportunity. Her lunches were spent discussing issues with others in our common sitting room. She was not afraid to ask help from people in the department where she sought the best advice. She learnt a great deal and others learnt from Ping, especially about new perspectives and points of view on their research both in the lab and in the field.

Recommendations

There is never enough time, but deadlines provide motivation for completion. Ping was able to complete a number of quality projects. Nine months may seem short, but this forces the labs and artists to focus on an agenda of completion. Perhaps a full year would allow more flexibility, but Ping was able to accomplish a great deal in the time provided. Some funds to help cover costs of consumables may help some artists in the future. Ping was fortunate, as Eawag covered the costs of consumables.

www.eawag.ch

THE BRAIN MIND INSTITUTE (BMI) | EPFL LAUSANNE

Focus

The mission of the Brain Mind Institute (BMI) is to understand the fundamental principles of brain function in health and disease, by using and developing unique experimental, theoretical, technological and computational approaches. We combine different levels of analysis of brain activity, so that cognitive functions can be understood as a manifestation of specific brain processes; specific brain processes as emerging from the collective activity of thousands of cells and synapses; synaptic and neuronal activity in turn as emerging properties of the biophysical and molecular mechanisms of cellular compartments. Understanding information processing in the brain and its higher emerging properties is arguably one of the major challenges in the life sciences.

Research at the BMI focuses on three main areas:
1. Molecular neurobiology and mechanisms of neurodegeneration.
2. Molecular and cellular mechanisms of synapse and microcircuit function up to the behavioural level and including metabolic aspects.
3. Sensory perception and cognition in humans.

Pierre Magistretti
Director of the Brain Mind Institute, Head of the Laboratory of Neuroenergetics and Cellular Dynamics
Olaf Blanke
Head of the Laboratory of Cognitive Neuroscience (LNCO)

We hosted an artist in order to promote exchange and stimulate a dialogue between the arts and sciences and to bridge the gap between the world of science and the general public through works of art. We compared the creative process in art and science, worked on projects with the artist, and discovered new and interesting questions for science and art. We also wanted to learn how a contemporary artist would benefit from an inside look at the activities of scientists.

Luca Forcucci was interested in the psychology of sound perception, memory and emotion, as well as their neural mechanisms. Obviously music and the brain was another interesting aspect for Luca. He learnt about neuro-imaging techniques that allow researchers to measure brain activity in behaving human subjects. He was especially fascinated by the recordings of electrical brain activity (electroencephalography, EEG) and how EEG experiments are designed. Luca further delved into topics such as how the body and self of the observer (of artworks) could be represented by the human brain. He was also interested in our virtual reality experiments and their application to cognitive science.

Luca mainly worked with Par Halje and Oliver Kannape, two of the PhD students at LNCO. Par, a physicist, introduced Luca to the principles of electroencephalography (EEG) and how these data are recorded and analysed. Par's research focuses on vestibular and own body perception and its neural correlates. Oliver's research focuses on action representation during walking and its importance for consciousness. Luca and Oliver discussed the LNCO's main areas of research ranging from multisensory corporeal perception to models of self-consciousness. Once Luca was able to grasp a common vocabulary for discussion, we then progressed to an in-depth exploration of those topics which he felt most intriguing and relevant to his project. This often meant moving away from the lab's ongoing projects to look at general principles of the (cognitive) neurosciences. Par assisted Luca in developing his own brain-imaging experiment in which he used EEG recordings to investigate the neural correlates of mental imagery and sound/music. The stimuli for the experiment were drawn from some of Luca's previous soundscapes and were used to evoke strong mental images. The analysis of this study is ongoing.

In his work *Music for Brain Waves* Luca experiments with online scalp recordings of the brain's electrical activity (EEG) and converts these measurements into a musical score or soundscape. This is made not just as a performance for an audience, but also as an input for the participant. This artistic setup has the inherent property of creating a closed loop between the music derived from brainwaves (the output of the participant's mental activity) and the music generated for brainwaves (the input). The initial stimulation evokes a 'reaction' from the participant, which in turn is 'sonarised' and fed back to him/her thereby enabling a fascinating dialogue or perspective-change for the participant as he/she switches back and forth between being the sound generator and the listener (the performer and the audience), incorporated or at an extra personal position. Luca created additional auditory stimuli for this experiment by converting bitmaps and videos of cellular activity into MIDI sound tracks. These are used during the performance to underscore and modulate the EEG-generated sounds.

Collaboration

Luca Forcucci and Olaf Blanke collaborated on an installation at EPFL's school of architecture (Archizoom's *Corps Sonore*). The sound installation integrated city and bodily sounds in the space of a room and distributed both sounds to specific locations within that same room.

Luca Forcucci, Isabella Pasqualini, Tej Tadi and Olaf Blanke also submitted a further installation project to an art festival in Montreal. Olaf Blanke, Luca Forcucci and Sebastian Dieguez collaborated on a short written article/comment that was published in Nature, December 24, 2009. The article was entitled *Don't forget the artists when studying the perception of art* and briefly discussed the role of the artist in the neuroscience of art. Luca Forcucci, Isabella Pasqualini, Tej Tadi and Olaf Blanke are finalizing an installation to be shown at the opening of EPFL's new learning center in spring 2010.

Effect on the Scientists?

Having Luca in the lab was very stimulating as it promoted dialogue not just between 'the artist' and 'the scientist' but between everyone in the lab. His unique viewpoints on and understanding of (bodily) perception were in many ways novel to us. To re-phrase our research findings and to find a common denominator with the artist proved especially educational as our long-established results were viewed and interpreted from a completely unfamiliar angle. Having Luca at LNCO and having discussion meetings on a biweekly basis also sparked several new research projects in what may be considered the 'cognitive neuroscience of art'. These discussions

Aerial photograph of the University of Lausanne and the EPFL Switzerland

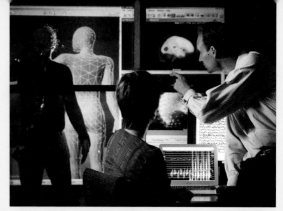

An analysis of out-of-body experiences and experimental laboratory illusions (Virtual Reality Lab, Cognitive Neuroscience EPFL), (Copyright: Olaf Blanke)

and projects included neuroscience and architecture, neuroscience and painting, the neuroscience of music, and the neuroscience of installation art.

Comparing Methodologies

At first both approaches may seem similar in that there is often an inspiration or novel idea, a *Geistesblitz*, which in the case of the artist may develop into a concept similar to the scientist's theory or hypothesis. In many ways though, art has more freedom to break with the norm in places. Science incrementally evolves from established methods and theories. All intermediate steps in a work of art shape its final form – rarely however, will demands be made from the artist to re-evaluate his initial vision. This is of course different from the natural sciences where unexpected data have to be investigated and understood. The scientist sometimes rethinks, reformulates or even discards his initial hypothesis.

Recommendations

The time period is too short to develop a more serious interaction. We propose to keep the common scheme of funding of 9 months, but to add a second more competitive artists-in-labs funding scheme to extend the artist-scientist collaboration with more specific follow-up projects. Perhaps in the future the artists-in-labs program could be open to art critics and art historians as well?

http://lnco.epfl.ch   http://lndc.epfl.ch

THE HUMAN COMPUTER INTERACTION LAB (HCI LAB)
INSTITUTE OF PSYCHOLOGY | UNIVERSITY OF BASEL
CENTER FOR COGNITIVE PSYCHOLOGY AND METHODOLOGY

Focus

The Faculty of Psychology has the distinction of being the youngest faculty of the University of Basel's seven faculties. It was founded in 2003, 25 years after the creation of an Institute of Psychology. This young faculty has quickly developed into a thriving hub for psychological research. Its research program, emanating from nine labs, focuses on human development and mental health, social cognition and decision-making, and neuroscience. Human Computer Interaction is a focus at the Center for Cognitive Psychology and Methodology. Human Computer Interaction Science (HCI) investigates processes, dialogues and actions of users who interact with a machine (often a computer) in a given setting using various methods such as: Interactive Prototyping, Expert Evaluation, Usability-Testing and Eyetracker-Analysis.

Sandra P. Roth
M.Sc., Assistant, PhD Student
Alexandre N. Tuch
M.Sc., Assistant, PhD Student
Peter Schmutz
M.Sc., Assistant, PhD Student

Alex Tuch is interested in the role of aesthetics in Human Computer Interaction, especially the way aesthetics affects the user's emotional and cognitive state during interaction. He also conducted studies on the effects of visual complexity on cognitive and emotional processing of websites within the framework of aesthetic theory and psychophysiological emotional research. Sandra P. Roth: The main focus of her research is users' expectations of web pages.

She explores user expectations regarding the location and existence of web objects and how arranging them in more or less expected ways influences objective usability and subjective ratings of different types of websites. Further, she is interested in how internet users perceive various web objects and how attention is best guided across important attributes of interfaces. The aim of her research is to improve usability by simple means and increase the overall user experience of websites and computer interfaces.

artist-in-lab

The aim of Monika Codourey was to investigate the cognitive and emotional patterns of constant travellers. Based on the acquired knowledge, she intended to develop a game prototype. Incorporating the understanding of this new nomadic lifestyle, her project perfectly matched in our research area of Cognitive Psychology and HCI. For her process we could provide her with a solid theoretical basis within Cognitive and Emotion Science and also give her appropriate research tools/instruments to assess behavioural, cognitive and emotional patterns of the constant travellers. To achieve her goal of how to identify the cognitive and behavioural patterns of constant travellers, Monika conducted a series of interviews with a previously developed interview manual. In this first step of a rather explorative nature she could identify important cognitive and behavioural factors of her subjects, which she later used to create an online survey. In the online survey Monika tried to confirm the finding from the interviews and generate empirical evidence from quantitative data. The survey was spread through social and business networking platforms and was presented at the ISEA

Observer room in the usability laboratory of the institute

DATA RECORD of mobile identities [5.min travel form]

Screenshot of Monika Codourey's project web page

2008 congress in Singapore <http://www.isea2008-singapore.org>. With the help of the scientists the gathered data was statistically analysed and prepared for presentation. The results from the interviews and the survey were then presented at the colloquium of the Center for Cognitive Psychology and Methodology to the staff members and Masters students. Furthermore, Monika conducted a usability test on the eye tracker. She analysed her web page <http://mobile-identities.info> and the related online survey. By doing so she could observe how the test participants perceived her web page and especially how they behaved on the constant traveller online survey. She used those insights to improve the quality of the survey and optimise its usability.

Effect

It is difficult to estimate in what way the scientific community (members of our department) has been affected by the artist. We can only report that researchers who were directly involved in the project were activated to reflect upon their own work by the artist's presence. By explaining our scientific methods and the way of how we investigate, the artist often confronted us with critical comments and made us reflect on our own methods. The approach of the artist was less structured and analytical than what we

are used to. The major discrepancy between the artist and the scientists was the different understanding of the term 'doing research on something'. However, by several hours of intense dialogue we could build up a common understanding about each other's approach and start working together on the same aim.

Recommendations

We think that a longer time period of about 12 months would be better to pursue the goal of concluding the project. The reason for this is that about 3 months are needed to get to some kind of common ground to understanding each others approaches towards research.

www.psycho.unibas.ch/mmi

THE PHYSICS DEPARTMENT AT THE UNIVERSITY OF GENEVA | CERN

Focus

The Department has about 300 collaborators and follows four main research directions, each represented by an institute: Condensed matter physics, Particle physics, Theoretical physics and Applied physics.

There are numerous collaborations between researchers in these directions, and with many partners outside the physics department, like the Geneva observatory, CERN, ESA and NASA. For details see <http://www.unige.ch/sciences/physique/index.html>.

Martin Pohl
Head of Physics Department

The Science

Christian Gonzenbach is an artist, but he is also a scientist because he has a very thorough approach to learning scientific facts. He did not learn through formulae and formalisms, but through discussion, reading, thinking, questions, answers, repetitions and corrections. This method compares to a 'normal' science curriculum and is similar to the total immersion method of learning a language in school. It is erratic, but efficient, based on trial and error, and, most importantly, it does not use mathematics, it uses people. Gonzenbach has been mainly interested in quantum phenomena at large. His collaboration with researchers studying the influence of quantum phenomena on larger scale systems and processes was particularly intense. In addition, cosmology, astrophysics and particle physics fascinated him, and he studied these subjects with remarkable attention. The subjects that modern physics focuses on can be globally categorized as:

- Matter: Its building blocks at the quantum level can be ordered into a periodic table of quarks and leptons, as well as their aggregates. It is the mission of physics to fully understand these ingredients and measure their properties.
- Forces: The typology of forces is studied with the aim to exhaustively identify them, measure their properties, intensities and distance laws.
- Space-time: The most mysterious ingredient is the vacuum, with its (apparently) three spatial and one temporal dimensions. It has long been seen as an empty stage, where matter and forces act on each other. It appears more and more obvious in modern physics that space-time itself takes an active role in this interplay; it may even well be the director of the piece.

The approach of physicists to these subjects is entirely analytical. The ingredients must be exhaustively listed, all their properties measured to the highest possible accuracy, and their different roles fully understood. In this way, phenomena become calculable and predictable, if not in a deterministic, then in a statistical manner.

The language we use in modern physics is highly imaginative, and often full of caustic humour: quarks, black holes, big bang, dark matter... Some of these terms have even been invented by adversaries to ridicule a discovery. Scientists have long forgotten the metaphoric origin of these denominations, for them these absurd names describe existing phenomena or objects of nature. Art can work with the cultural connotations of this absurdity. The size of our subjects of study is often equally absurd, absurdly small or

The AMS space experiment in its final phase of completion, to be deployed on the International Space Station in 2010.

A view of Christian Gonzenbach's atelier at the Physics Department of the University of Geneva

absurdly large. Since these sizes defy imagination, we take rescue to imperfect mental images or models. And we remind ourselves about their imperfection constantly, since the only truth is in the mathematical description. The artist is not bound to these rules. He/she can take the models literally or metaphorically, as best suits his/her goal.

The artist-in-lab

The questions we initially hoped the artists-in-labs residency would help us to approach were the following:

- Is there a non-analytical approach to our subjects of study? Can one see, feel, or hear science? Can one smile or laugh about it?
- Is there a non-mathematical way to communicate science? How do we talk about it? How should we talk about it?
- And most importantly: Is there a new way to see the scientific process? How do we do science? What is important in the process and what is not?

When one proposes a scientific experiment, one anticipates its result. Researchers push the limit of the known and measurable, but without leaving solid ground. And experts mandated by our funding agencies judge the proposal based on their own experience, and often enough on their own failures. But it is in fact rather rare that an experiment has the anticipated outcome.

The most interesting experiments are the ones that 'fail'. Not in the sense that they do not work, but in the sense that they yield new, unexpected phenomena. Instead of adding concrete to the foundations of current wisdom, they crack them. One of the reminders that we owe to the artists-in-labs residency is that we should be more playful and bold in proposing experiments; that we should not be afraid of failure or uncertain outcome of what we do.

What was beneficial and interesting, was to observe an artist play with the scientific method. He was at liberty to modify it in any way he pleased, make fun of it, even pervert it. That way he shed a new light on the scientific process itself, allowing scientists to become again conscious of the way that science is done. And to become conscious of the fact that sometimes the process is more important than the result. For us, the outcome of the artists-in-labs residency has thus been not to create knowledge, but to comprehend and question the scientific process itself. Christian Gonzenbach's project was to do the impossible, thus it could only fail. He wanted to give shape to the untouchable, shed light on the invisible, watch empty space replenish, reconstruct atomized matter, map out dimensions beyond the four we know. He still has a number of projects pinned to his atelier wall. Some very concrete, some completely impossible, some as absurd as the words we use. None of them is illustrative in any way. Some make reference to his prior work, most are completely new. All of them mean something, but not the same to him and to us and so he gives an extra dimension to our science. Through his work he brings out the symbolic dimension in physics. And that is a major discovery. And so we commissioned a QUantum ARt Chamber *QUARC*: a cabinet of curiosities, in which he will exhibit sculptures, kinetic objects, movies and other products of his residency. The *QUARC* was installed in the entrance hall of the Physics Building at the University of Geneva, and it will stay as long as both the artist and the public are interested in keeping it there.

Conclusions

The artists-in-labs residency did not create scientific knowledge, but sponsored an artistic process and the creation of art out of its usual context. For us scientists, artists-in-labs has made a UFO land on our premises. The residency has contributed to a renewed reflection about scientific and artistic processes, with respect to other creative processes. A symbolic dimension to physics has emerged in the minds of the participants. Artists-in-labs do not change science, but the scientists involved. One may hope that it also changes the artists involved. In any case, the *QUARC* will stay as a quasi-permanent meeting place for Martian artists and Earthling physicists.

www.unige.ch/sciences/physique/index.html

PAUL SCHERRER INSTITUTE (PSI) VILLIGEN

Focus

The Paul Scherrer Institute (PSI) is a multidisciplinary research centre for natural sciences and technology. PSI collaborates with national and international universities, other research institutions and industry in the areas of solid-state research and materials sciences, particle physics, life sciences, energy research and environmental research. PSI concentrates on basic and applied research, particularly in those fields which are the leading edge of scientific knowledge, but also contributes to the training of the next generation and paves the way for sustainable development of society and economy. The institute is actively involved in the transfer of new discoveries into industry, and offers its services as an international centre of competence, to external organisations. PSI employs 1200 staff members, making it one of the largest Swiss research institutions. It develops, builds and operates complex research facilities that impose particularly high requirements in terms of knowledge, experience and professionalism. PSI is one of the world's leading user laboratories for the national and international physics community and hosts 1500 visiting scientists per year.

Beat Gerber
Communication Officer in Staff of Director of PSI
Fritz Gassmann
Senior Researcher, Staff-Member of General Energy Department
Robert Maag
Responsible Engineer for Deconstruction of Nuclear Test Reactor SAPHIR
Marcel Dänzer
Head of Technical Lab 'Lehrlings-Werkstatt'

During the artists-in-labs residency, Beat Gerber was the Communication Officer for PSI. He was responsible for correct and comprehensive information adapted to a form understandable by a large public. Main outputs are the annual report of the Institute (progress report), press releases for important results, breakthroughs, and events. Fritz Gassmann was a Senior Scientist (physicist) contributing to different projects such as climate change research, air pollution, market diffusion of innovative technology, *Energy-Globe* in the visitors centre *(PSI forum)*, all projects giving an overview on energy research at PSI. His special research field was nonlinear dynamics, chaos, and complex systems. During 2007, he was building a students' lab called *iLab* <http://www.ilab-psi.ch> for young people to show them the fascinating sides of physics and to interest them in the natural sciences.

artist-in-lab

With the initial title *Energy Plan for the Western Man*, Roman Keller attempted to develop a technical piece of art that would illustrate a PSI-project in a way scientists would never try. His ideas focussed on energy from biomass to drive cars. Keller imagined a kind of children's car that would be driven by burning flowers. He introduced many ideas to the scientists of the general energy department in a seminar (March 8). After many discussions, this idea had to be abandoned because its realization turned out to be too complicated, but he constructed a new vision: The world's first solar rocket. Based on research made about steam rockets at the beginning of the 20th century and on experience found in connection with a Swiss rocket club, a rocket like this could be

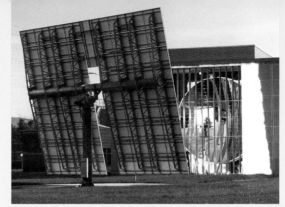

The PSI is the largest national research centre in Switzerland and a renowned international user lab (Photo: PSI)

Solar furnace of PSI test facility for high temperature solar technology (Photo: PSI)

expected to reach 400–700 m. It could be worked out in a short time, but the realization of the system proved much harder than he imagined. It was not simple to find a bottle which resisted a pressure of 15 bars that could be heated in the focus of a parabolic mirror. Furthermore to connect and construct a nozzle with the chosen SIGG aluminium bottle was a challenge and more time than expected had to be devoted to find a solution. Components were tested with electric heaters to replace the fading autumn sun. However, with hard work, a very nice rocket was finished in December and Roman Keller arranged a little exhibition in a container showing it together with an impressive reflector and different tools and pictures illustrating the various stages of its development. Keller also presented his book *The Rocket for the Rest of Us,* where he documented the development of hot water rockets and of solar energy from earlier times (Archimedes) and ending with his solar rocket. Many astonishing pictures make this book interesting and fascinating to read! Parallel to the development of the rocket, Roman's lecture was presented in the June issue of the PSI inhouse journal SPECTRUM, for which he also contributed a series of photographs of PSI-buildings. This showed the scientists and how they work in them. In the December issue of SPECTRUM, he featured another article with the first 10 m test flight of the solar rocket. In September, Keller gave a second presentation in PSI forum on his work and its implications for energy source.

Recommendations

We were astonished that an artist wanted to construct a technical system that was fully functional. As it is the case with all our developments of technical systems, the devil is in the details and they need A LOT of time. We think that such a technical development should be distributed over a longer period. Would it be possible that the artist visits the institute only one day per week during the construction period so that his 9 months are spread over a longer time? He will demonstrate his rocket to interested PSI-scientists in the next years.

www.psi.ch

CSEM SWISS CENTER FOR ELECTRONICS AND MICROTECHNOLOGY ALPNACH

Focus

The CSEM Swiss Center for Electronics and Micro-technology Inc. (CSEM), based in Neuchâtel, Switzerland, is an established research organization active in the fields of micro- and nanotechnology, microelectronics, robotics, photonics and communication technologies. The company sees itself as a bridge between scientific research and commercially viable results for industry and this approach is characterized by:

1. Development of key technologies
2. Integration of key technologies into innovative products
3. Acquiring cutting-edge know-how by applied research
4. Developing core competencies to enhance customers' competitive advantages
5. Boost regional business development

CSEM's aim is to conduct applied research and development for the micro- and nanotechnology industries. This is coupled with the idea of technology transfer to industrial partners and setting up spin-off or start-up companies. In fact, during the past eight years, CSEM has founded 29 new companies.

Approximately 400 highly qualified and specialized employees from various scientific and technical disciplines work for CSEM in Neuchâtel and the four regional centers in Zurich, Basel, Alpnach and Landquart. They represent more than 30 nationalities and constitute the basis of the company's creativity, dynamism and innovation potential. Hundreds of new jobs have been created in its spin-offs and start-ups.

Swiss, European and Worldwide Partnerships

Providing interdisciplinary technology-based solutions requires more and more competencies which cannot be found in a single organization. Therefore, national and international partnerships between technology providers are essential. This is a universal fact, but it is of particular importance for Switzerland, which is a small country, with limited resources, albeit with a dense network of high-quality academic institutions and R&D centers. CSEM also maintains privileged relationships with the main Swiss research centers such as EMPA – a materials science and technology research institution – and the Paul Scherrer Institut (PSI), as well as through the Swiss-French platform MIND, dedicated to microelectronics and mechatronics.

The international nature of business today makes the extension of partnerships beyond national borders of paramount importance. Partnerships with complementary R&D organizations allow a broadening of the technology offer, and therefore increased relevance and appeal for the industrial customer. As an example, CEA (France), CSEM (CH), Fraunhofer Gesellschaft (Germany) and VTT (Finland) formed the Heterogeneous Technology Alliance (HTA), to face the challenges of micro- and nanotechnology transfer. These organizations also created a commercial unit, the company 4-Labs, to exploit potential synergies.

Philippe Steiert
VP, Regional Centers
Dirk Fengels
Section head Sensors & Systems

artist-in-lab

The engineers and researchers of the Alpnach division develop and provide innovative solutions in the fields of Microassembly and Robotics, Microfluidics and Microhandling, Sensors and System as well as Optics and Packaging. These foci are applied research results for industrial customer projects. Inspired by our electromechanical systems and robots in the lab, Pe Lang chose to combine his passion of audio engineering with precision motor drives. After collecting many ideas and impressions in the lab, one dominant project idea evolved. The artist's vision was a highly aesthetic, dynamic speaker system that plays with sound and motion in space, suggesting an interactive dialogue with the observer. The work consisted of designing the speaker platform, choosing or designing appropriate system components such as speakers, motors, amplifiers, controllers, encoders and making interconnections. The artwork included several speakers that could be controlled by a software platform on a computer, programmed by Pe Lang. He worked towards a configurative software system, capable of playing programmed choreographies of sound and motion. One intention of the project was to capture an observer by audible and visual impressions, suggesting a dialogue between artwork and observer. Although several ideas sparked during the initial search, 9 months only allowed him to implement the project described above. While the main electromechanical system could be constructed within 6 months, the control software could only be realized to a state of basic functionality and the synchronization of sound and motion. Remaining tasks included the software means for implementing choreographies. Due to this time factor, the speakers

were presented to the institute. Having an artist in the lab triggered discussions about perceived value and the role of aesthetics in technology. It encouraged different paradigms, beyond the existing diversity of perspectives inherent inside this highly multidisciplinary community. For some engineers and scientists, the expected gap between art and science became much smaller. Others discovered value in spending more thought on aesthetics when it came to creating attractive technology demonstrators, possibly leading to an increase in the attention of potential customers. The artistic approach was surprisingly similar to the methodologies applied in innovative engineering because Pe has a professional technical background and his tools and working materials were similar to the ones used by the scientific institute. On a more abstract level it can be claimed that insights gained from observations and exchange within an interdisciplinary environment and combined with own experiences, skills and paradigms, lead to the development of a system that was never seen before and therefore truly innovative. However, while most scientific and engineering approaches focus on results, predetermined by early specifications to a large degree, the artistic approach seems to emphasize an evolutionary process rather than focus on the search for a method to solve an existing problem or increase performance.

Recommendations

The amount of time was adequate, however, it was too short to permit entire completion of the project. It provided a sound scope for the artist's evolutionary process and prototyping. It was important that Pe Lang was able to complete the project to his satisfac-

CSEM's headquarters in Neuchâtel – main building

CSEM miniature robots for microassembly (PocketDelta)

tion within reasonable time, and we really enjoyed the cooperation and the residency. In retrospect we missed the opportunity to encourage the artist to exhibit his work in the lobby or to give more presentations of his work to those interested. At the same time, scientists and engineers are likely to be under severe time and budget constraints. This did limit the amount of time they could spend with the artist. Perhaps some tips about how to deal best with this situation may help us in the future.

www.csem.ch

NATIVE SYSTEMS GROUP | COMPUTER SYSTEMS INSTITUTE ETH ZURICH

Focus

The main goals of the institute are to conduct research and development in programming languages and compilers, custom run-time systems, dependable software for safety critical applications, pervasive computing and Human Computer Interaction and finally innovative applications in eHealth and the Digital Arts. We attempt to develop powerful and widely usable IT-based tools for the Digital Arts. We already have a history of experimenting with new programming paradigms and in particular with live-coding and of working with musicians, visual artists and dancers.

Jürg Gutknecht
Chairman Department of Computer Science
Sven Stauber
Doctoral student Native Systems
Felix Friedrich
Researcher & Lecturer Native Systems
Nicholas Matsakis
Laboratory for Software Technology
Christoph Angerer
Laboratory for Software Technology

We consider IT as an important and powerful tool platform for artists to realize and enhance their creations. While a myriad of highly useful authoring, editing and synthesizing tools starting with Photoshop, Director and Flash and ending with Max/MSP (to name just a random selection) have raised both the variety and the quality level of artistic multimedia creations to new heights, these systems are often cumbersome for use by artists, and they are sub-optimal in many cases for the specific purpose envisioned. One obvious way out of the dilemma of general purpose software off-the-shelf vs. user friendliness for some specific application is building custom software. Given this background, and making use of the well-developed programming skills of our artist-in-lab, we have tried to explore the custom software approach at the example of a composer system for audio/video productions based on the metaphor of object-oriented programming (in contrast to an interactive, graphical approach). The result was an audio/video programming system based on the *Self* programming language. Thanks to its dynamic object model, live coding is possible in the sense that properties and functionality can be assigned dynamically to objects during performance time.

artist-in-lab

The artist, Chandrasekhar Ramakrishnan developed a novel programming environment that allows artists to realize their own multimedia performances. The entire duration of the residency was spent on this project. The artist presented some of the ideas behind this platform during one of the weekly institute seminars and at our yearly retreat. An excerpt from the performance piece was shown at the artists-in-labs conference *Trespassing Allowed* in December 2008 at the ZHdK.

The project raised the awareness of the potential to apply IT principles and methods to non-standard fields. It exemplified the power of software-controlled custom-built tools for making creative work and introduced the paradigm of live-coding. Chandrasekhar's approach was very pragmatic and oriented towards the intended purpose.

The Native Systems Group checking out parts of a running robot, 2008

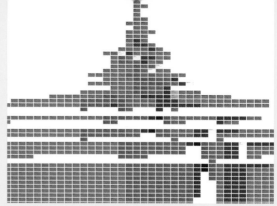

Modular design visualized at the example of the *Aos* operating system (Native Systems Group, ETH Zurich)

Recommendations for the Future

The time frame was at the lower limit and so we extended it and funded the artist to stay with us another 8 months in order to finish his platform.

http://nativesystems.inf.ethz.ch

ISTITUTO DALLE MOLLE DI STUDI SULL'INTELLIGENZA ARTIFICIALE (IDSIA) | USI-SUPSI

MANNO-LUGANO

Focus

IDSIA (Istituto Dalle Molle di Studi sull'Intelligenza Artificiale) is a non-profit oriented research institute for Artificial Intelligence, affiliated with both the Faculty of Informatics of the University of Lugano and the Department of Innovative Technologies of SUPSI, the University of Applied Sciences of Southern Switzerland.

Luca Maria Gambardella
director
Jürgen Schmidhuber
director

We focus on machine learning (artificial neural networks, reinforcement learning), optimal universal artificial intelligence and optimal rational agents, operations research, complexity theory, and robotics. IDSIA is situated near Lugano, a lakeside city in the Italian-speaking canton of Ticino, a region of Switzerland. IDSIA is small but visible, competitive, and influential. For example, its Ant Colony Optimization Algorithms broke numerous benchmark records and are now widely used in industry for routing, logistics etc. (today entire conferences specialize on Artificial Ants). IDSIA is also the origin of the first mathematical theory of optimal Universal Artificial Intelligence and self-referential Universal Problem Solvers (previous work) in general. Once AI was dominated by heuristics, but IDSIA's artificial Recurrent Neural Networks learn to solve numerous previous unlearnable sequence processing tasks through gradient descent, Artificial Evolution and other methods. Research topics also include complexity and generalization issues, unsupervised learning and informa-

tion theory, forecasting, learning robots, robots with artificial curiosity, robotic swarms, etc. IDSIA decided to participate to the artist-in-lab initiative to evaluate whether it is possible to influence art with our research activities and whether it is possible to take inspiration from art to improve our work.

artist-in-lab:
Alina Mnatsakanian

Alina was inspired by our research in robotics and the social insect metaphor. This research emphasises aspects such as decentralisation of control, limited communication abilities among robots, use of local information, emergence of global behaviour, and robustness. Most current studies in robotic systems have focused on robotic swarms in which individuals are physically and behaviourally undifferentiated. Therefore the main competencies of people involved in her project were artificial intelligence and swarm robotic researchers. We first asked Alina to follow the course at the University of Lugano concerning heuristics algorithms and to talk with all researchers at IDSIA. Then, Alina was exposed to the main artificial intelligence methodologies like knowledge representation, neural networks, genetic algorithms and ant colony optimization. We also taught Alina how to program a robot and how to use a dedicated robotic simulation environment. We helped her to develop the swarm dance choreography and the associated online movie. Alina worked on a very challenging project: her goal was to have a swarm of robots dancing together so we gave her a group of mobile robots that were able to move and to produce light effects. Alina programmed a nice choreography based on a sequence of sounds and all the associated

Entrance to IDSIA which houses directors, administration, senior researchers, researchers and PhD students

robots' movements. Her robots started with a synchronization phase and later on they danced together following minimal music. During the dance they also interact by using an obstacle avoidance mechanism that produces unexpected effects. The entire dance is seen by a video-camera and the images produced during the dance are processed and an artistic video is immediately visualized using a projector on the wall. This combination of real robot interaction, music and video produced a very interesting result that involves different technologies. The project took the entire nine months, as it was refined many times during profitable discussions with people at IDSIA. She really understood the robots' capabilities by the end of her residency, and made a series of public presentations of her artistic production to the entire institute. The first presentation was organized at the beginning to present Alina's previous work and her context, the other two were organized to show the current state of the project and to get feedback from all the IDSIA researchers.

Her Effect on Us

Alina was a very interesting person. She liked to talk about art and she stimulated our researchers into looking at what they are making from a different perspective. Some researchers started to see artistic potentials in their own research, for example, would it be possible to see if their dynamic processes, like neural networks learning phases or ant colony optimization pheromone updating, could become part of an artistic production? Some other researchers are investigating the notion of low-complexity art in order to try to represent the essence of objects in a very compact mathematical form. Scientific method-

ology is based on an empirical process of discovery and demonstration that starts from the observation of phenomena. Next a hypothesis about the phenomena is formulated and a set of experiments and demonstrations are used to validate or falsify the hypothesis. Conclusions of this phase are used to eventually iterate the process with other hypothesis, other experiments and demonstrations. In the case of Alina the goal was not to prove any theorem or to have a valid theory. However, she spent time in formulating a hypothesis about the dancing process and to validate the relation of using a real swarm of robots. In a more scientific approach the evaluation phase also requires an artistic component that is very difficult to formalize and to measure.

Recommendations

We consider this to be a really positive experience. We think that the length of the residency is sufficient to transfer knowledge from the lab to the artist and to see some artistic production at the end. Perhaps additional funding can be sought to support the realization of the final product!

www.idsia.ch

Focus

The main goal of Artificial Intelligence (AI) is to work out the principles underlying intelligent behaviour. These principles enable us on the one hand to understand natural forms of intelligence (humans, animals), and on the other to design and build intelligent systems (computer programs, robots, other artefacts) for research and application purposes. In addition, they change the way we view our bodies and the world around us. Our laboratory is highly interdisciplinary and international, attracting researchers, designers, and artists from all over the world. The overall research topic that we have been pursuing concerns the implications of embodiment. How can our brain, body, and environment interact to produce coherent behaviour? The overall 'philosophy' of our research program is provided in the book *Understanding Intelligence* (by Pfeifer and Scheier, 1999, MIT Press), and in *How the Body shapes the Way we think – a new View of Intelligence* (by Rolf Pfeifer and Josh Bongard, MIT Press, 2007).

Daniel Bisig
Senior Researcher

Our research program consists of nine major research strands that all contribute to the overall goal of Artificial Intelligence: Locomotion (with a focus on morphology, material properties and body dynamics), Development and Learning (an area which has become known as 'developmental robotics'), Evolution and Morphogenesis (in particular the co-evolution of morphology, materials, and neural substrate), Collective Intelligence (in particular the emergence of spatial and social patterns), Art and Design (which includes interactive and generative biological simulations and social robotics), Neural Interfacing (the combination of artificial and natural neural systems), Self-Organization in Modular Systems (in particular self-assembly, self-reconfiguration and self-repair), and finally Artificial Physicochemical Systems (towards the design of an artificial cell).

Our program is based on the conviction that:

a) intelligence is a multifaceted research field that requires the investigation of a wide range of topic areas, and

b) the interaction between the various disciplines is highly productive.

For example, engineers and computer scientists can learn from nature, that is, from animals and natural evolution, and biologists and psychologists can learn from building robots and developing computer programs. In our laboratory researchers from a large diversity of backgrounds such as computer science, mathematics, physics, biology, ethnology, neuroscience, psychology, philosophy, mechanical and electronic engineering are cooperating on a number of projects towards our overall goal of understanding intelligence.

We have already hosted 3 artists-in-labs, and a number of artists have worked in our laboratory for a few months, an experience that was highly beneficial for both scientists and artists. All projects contribute, one way or other, to our central theme. More specifically, they are intended to explore the core concepts that form the basis of our approach: they include embodiment (the physical realization of agents), morphology, system-environment coupling, dynamics, and material properties.

Various robotic investigations in the Artificial Intelligence Lab, University of Zurich, 2008

Group photo depicting Rolf Pfeifer, the head of the AI Lab, and most of his senior researchers and PhD students

Our artist-in-lab: Pablo Ventura

For the duration of nine months, the choreographer Pablo Ventura collaborated with scientists at our laboratory on a project entitled *Choreographic Machine*. Pablo learnt about Artificial Life Simulations and how to integrate these simulations into a choreographic process that employs the *Life Forms* software in order to create a fully automated computer based choreographic software. The main audacious goal of this residency was the development of a human sized robotic dancer that would be controlled by this software. This project came from an initial proposal by Pablo Ventura and represents a combination of the artists' and scientists' interests. In particular, the project reflected Pablo Ventura's fascination with algorithmic and automated processes as a means to free choreography from cultural traditions. Furthermore, the project attempted to blur the distinction of machine and human characteristics and thereby challenge the spectator's own self-image. From the scientists' point of view, the project promised to produce valuable insights into principles of self-organization that can lead to the emergence of universal (natural and cultural) movement patterns and helps in the establishment of bio-mimetic design principles for the creation of natural movements in a humanoid robotic system. However, the duration and costs of the project clearly exceeded the funding available. Members of the lab are currently applying for additional funding to continue the project. The artists-in-labs program has played a key role in the initiation of this long term collaboration.

Recommendations

The time and costs of large projects that involve the development of both hard and software components for a robot clearly exceed the scope of the artists-in-labs program. For this reason, it would be good if the artists-in-labs program could become accepted into a network of related science funding initiatives that may help to support more long-term collaborations costs in the future.

http://ailab.ifi.uzh.ch

123

ARTISTS IN RESIDENCE FINAL REPORTS

Life Sciences:

HINA STRÜVER & MÄTTI WÜTHRICH
ARTISTS IN THE INSTITUTE OF INTEGRATIVE BIOLOGY (IBZ) | ETH ZURICH

SYLVIA HOSTETTLER
ARTIST IN THE CENTER FOR INTEGRATIVE GENOMICS (CIG) | UNIVERSITY OF LAUSANNE

CLAUDIA TOLUSSO
ARTIST IN THE WSL SWISS FEDERAL INSTITUTE FOR FOREST, SNOW AND LANDSCAPE
RESEARCH, BELLINZONA

PING QIU
ARTIST IN THE EAWAG: THE SWISS FEDERAL INSTITUTE OF AQUATIC SCIENCE AND
TECHNOLOGY, DÜBENDORF

Cognition & Physics:

LUCA FORCUCCI
ARTIST IN THE BRAIN MIND INSTITUTE (BMI) | EPFL, LAUSANNE

MONIKA CODOUREY
ARTIST IN THE HUMAN COMPUTER INTERACTION LAB (HCI LAB) | INSTITUTE OF PSYCHOLOGY
UNIVERSITY OF BASEL

CHRISTIAN GONZENBACH
ARTIST IN THE PHYSICS DEPARTMENT AT THE UNIVERSITY OF GENEVA | CERN

ROMAN KELLER
ARTIST IN THE PAUL SCHERRER INSTITUTE (PSI), VILLIGEN

Computing & Engineering:

PE LANG
ARTIST IN THE CSEM SWISS CENTER FOR ELECTRONICS AND MICROTECHNOLOGY, ALPNACH

CHANDRASEKHAR RAMAKRISHNAN
ARTIST IN THE NATIVE SYSTEMS GROUP | COMPUTER SYSTEMS INSTITUTE | ETH ZURICH

ALINA MNATSAKANIAN
ARTIST IN THE ISTITUTO DALLE MOLLE DI STUDI SULL'INTELLIGENZA ARTIFICIALE (IDSIA),
MANNO-LUGANO

PABLO VENTURA
ARTIST IN THE ARTIFICIAL INTELLIGENCE LABORATORY | UNIVERSITY OF ZURICH

Fragile balance: An artificial plant is growing. Performance *Soil,* Switzerland (Photo: Nara Pfister)

Hina Strüver & Mätti Wüthrich

ARTISTS IN THE INSTITUTE OF INTEGRATIVE BIOLOGY (IBZ) | ETH ZURICH

Disciplines: Performance Art, Installation

Project title: *Regrowing Eden.* Ever since human beings were expelled from the Garden of Eden, the longing for paradise remained. Gene technology seems to be a possibility to rebuild the Garden of Eden on Earth. For some, GMOs are a way to paradise on Earth, for others they are just another doom of temptation. The project attempted to make a performative and artistic mapping of the actual social and ethical discourse of the GMO Research at the Institute of Integrative Biology.

REGROWING EDEN Hina Strüver & Mätti Wüthrich

Artists outside the lab (May–June)

When we initially applied for the artists-in-labs program we were overwhelmed by a sudden feeling that we needed to broaden up our scope and to get out of the lab and into the public. While the scientists in this lab study and verify the risks of genetic engineering (GE) particularly the risks of planting genetically modified organisms (GMOs), these crops have already been planted out there in the real world. Alongside their investigation on the ground, we wanted to explore how scientists and policy makers communicate the risks of GE/GMO to the public and how the public perceive these risks. Therefore, we decided to spend time in Switzerland, partly working in the lab, but also to go to Brazil and Vietnam in order to meet other scientists and stakeholders and to make reactive installations and performances at these sites. These three main countries were chosen because of their relevant attitudes towards GE. We could build on the Institute of Integrated Biology GE/GMO risk assessment programmes under the direction of Dr. Angelika Hilbeck, our scientific partner. We started our residency by learning some GE-Basics. We extracted DNA and analysed their activity with a photo spectrometer and we learnt how to design a proper risk assessment and apply those scientific methodologies. For example how to measure unwanted or negative side effects like the effects of GMO on non-target organisms or on the nutrients in the soil fertility. However, in order to capture the opinion of the scientists in this lab and to gain an impression of the social discourses taking place on GE/GMO we developed a standard sociological questionnaire and made interviews with the scientists. We also interviewed policy makers, artists and any other stakeholders that we could find. GE/GMO modification is complex and very difficult to understand. We tried to involve as many people as possible and placed a great emphasis on the communication of the controversy itself. Therefore, right in the beginning we created a website called: <http://regrowingeden.ch>. It is basically an accessible communication platform with the project-specifications, an online-questionnaire with feedback tools, a blog, the photo documentation of the performances that took place, as well as the visual representations of the 'virtual garden edens' calculated from the questionnaires. Slowly over the course of the next nine months we translated all content into German, English, Portuguese and Vietnamese.

Our main methodology was simple and effective: travel to these countries and get a grip of the local situation by talking to the relevant scientists, policy makers, artists and general public. In this way we gained a specific impression of the GE/GMO situation in each country and made performances. Then we provided the opportunities for feedbacks and interdisciplinary discussions through self-organised art cafés, parties or art & science dinners.

regrowingeden.ch (Switzerland, June–July)

When we saw the bright green floor of the giant inner courtyard of the Institute of Integrative Biology, we knew that this would be the place for our first installation. The roof of the courtyard was topped with glass, where the light could enter and behind this roof, one could see the facade of the tower of the Department for Environment and Climatology.

Mapping the social discourse about GE/GMO on <http://regrowingeden.ch>

Mätti Wüthrich and Hina Strüver after a performance

We realized we could use our rock climbing skills as part of the performance in the courtyard and also outside on the roof. The small offices and the labs of the scientists all look out onto this courtyard, and for us they resembled cells of a bigger entity, like a living organism.

After a non-bureaucratic procedure we actually got the permission to make a one month performative installation in the building. In the preparation of the performances we climbed over these cells or windows and put up a net-structure or basic matrix for an artificial plant to grow within. Over time the installation grew by making 5 performance-interventions, including 2 climbing performances and 1 on the fragile glass roof. The result simulated the whole life cycle of a GMO-plant and became a reflection on how a GMO-organism might feel when bombarded with golden particles adhering foreign DNA in the process of the scientific DNA-intrusion. This should result in the creation of a new super species – the plant becomes a hybrid with functions from more than one species. In the final stage of our performative installation the GMO plant invaded the 30 m tall outer façade of the ETH-tower itself, escaping from the secured inner courtyard. Alongside this project, we organized art cafés for the scientists during their coffee breaks

and after the performances. Indeed, it was hard to distract the scientists from their research, even when we climbed by their desks on the outside of their windows!

Transgênicos (Brazil, August – September)

Brazil is a leading exporter of GMO-cotton and known for the destruction of rainforests in order to grow GM crops. The ETH gave us their scientific contacts, but we also engaged with other stakeholders like Greenpeace. According to Greenpeace Brazil, the hottest GMO-debate seemed to be happening in the state of Paranà, where a legal battle between the state and federal government has been going on for many years. So we went to the provincial capital, Curitiba, and linked up with the responsible state officials. They endorsed our project and helped us to organize a performance-installation at the famous Museo Oscar Niemeyer in Curitiba. The installation grew through three performance-interventions; slowly, a dense plant sculpture evolved. After the governors briefing, over 200 state officials as well as other invited guests attended one of our performances. We insisted that we were not political instruments nor 'artists against GMOs', therefore we made our artistic and scientific positions clear, wrote the

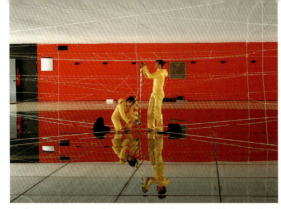

Performative installation: a plant will grow within a defined matrix, Museo Oscar Niemeyer, Curitiba, Brazil (Photo: Juliana Burigo)

Invasive growth: Details from the performance *Transgênicos* (Photo: Juliana Burigo)

statements ourselves and spoke to the press directly. We also organized an in-depth discussion about environmental issues in relation to art performance for the curators of the Oscar Niemeyer museum. We provoked great discussions about form and content of our art. Surprisingly performance as an independent form of art was a new concept to many visitors and even for the curators!

Viet Nam Eden (Vietnam, October-November)

When we landed in Hanoi – the city of the soaring dragon – we thought that such a performance about the debate of GMO crops would be challenging. The Socialist Republic of Vietnam is characterised by a system of political control and an ultraliberal economy and it is the world's biggest rice exporter. Therefore, the development of GE/GMO is a national top priority, despite the fact that there is no legal framework for any GE/GMO assessment actually in place! Would our artwork be censored, cancelled or, following the official political line, would we be seen and used as 'GE/GMO-promoters'? Anyway, we were welcomed and briefed by different Vietnamese GMO-scientists and policymakers. After two weeks we had an overview of the GE/GMO-situation in Vietnam. Consequently we contacted young artists and cura-

tors from the emerging and very lively performance art scene who proved to be very open to an exchange of ideas on art and science!

Luckily, after some trial-and-error-attempts the avant-garde artist/curator of the well known, but informal Nha San artspace accommodated our project and encouraged us to make uncensored performances. The native stilt-house architecture and the calm atmosphere of this space made a stark contrast with the GE high-tech labs and the aggressive pro-GE official politics. In two performances we created an organic looking, yet artificial GMO plant of plastic tubes within a clean matrix made of strings. We sucked yellow and red colour through the transparent tubes and wove the GMO-plant growth into the courtyard. The GE-content, the debate of it, as well as our hybrid form of art was very unfamiliar to the public so we invited artists, scientists and policy makers to dinners where we explained the topic and discussed how to cross the line between artificial and natural, art and science, installation and performance.

Your opinion on GE/GEMO as an artistic picture

Our website <http://regrowingeden.ch> contains an interactive set of 3D virtual 'genetically modified

Viet Nam Eden at Nah San artspace, Hanoi (Photo: Le Ngoc Tung)

Hybrid: Organic growth by artificial material (Photo: Le Ngoc Tung)

organisms' whose animated growth is determined by various parameters from a questionnaire about different opinions in different countries. This work was realized with the help of a computer specialist (Felix Marthaler) and a collaborator from the Institute of Informatics (Hansrudi Noser). The fractal images are calculated by custom software based on the *Lindenmayer Systems* – a program that has parallels to universal sets of genetic code. Here the answers from the questionnaires are readily transformed into a unique 'virtual plant'. An animated greenhouse of all the plants provides a social discourse on gene technology or a virtual garden of eden!

Mission completed – lessons learnt

It was an interesting and intense 9 months! We learnt a lot about genetics and about attitudes towards the subject in diverse sociopolitical contexts. We realized that scientists are human beings and that scientific results are rather subjective findings, which often depend on the deep rooted ideology and motivation of each scientist. Through our performances, we made many contacts and met new friends. Many of them were rather critical about the potentials of Genetically Modified Organisms. However, the level of understanding and involvement about GMOs dif-

fered between the three countries. The courtyard foyer of the research building of Swiss ETH allowed us to have the most freedom of experimentation and expression.

Switzerland is well known for its differentiated scientific and public debates that resulted in the famous GE/GMO-moratorium. In Brazil, GMO is a politically hot topic – this heat is represented by the red-walled Museum Oscar Niemeyer, where we were encouraged to make performances, because the Governor himself is an anti-GMO policy maker. In the inofficial Nha San artspace in Vietnam, we actually found ourselves representing the unofficial existence of GE/GMO. At the time of our visit in Vietnam, there were no legal GE/GMO frameworks in place, even though they were already growing in many places!

Conclusions and recommendations

With gene technology, humans are designing new life forms. The artists-in-labs program gave us a real opportunity to examine the complexity of genetic engineering and to reflect upon the interaction between nature and society. Both the artists-in-labs and the ETH-lab contributed financially to our trips in order to make comparisons between societies in rela-

Virtual Eden: 'Your opinion on GMO as fractal image', animation from questionnaires (Image: Hansrudi Noser)

Mutating DNA: Climbing performance Cell, ETH Zurich, Switzerland (Photo: Rebecca Naldi)

tion to this interaction. We had the unique possibility to experiment both inside and outside the lab and collaborate with each other.

Credits

Angelika Hilbeck, Antonio Pietrobelli, Christof Sautter, Duc Nguyen Manh, Evelyne Underwood, Felix Marthaler, Georg Bauer, Hansrudi Noser, Huynh Thi Thu Huong, Nara Pfister, Nguyen Hong Son, Nguyen Van Tuat, Le Hien Luong, Margit Leisner, Octavio Camargo, Rebecca Naldi, Tran Luong, ETH Zurich, the entire Nha San Hanoi-crew, the staff of the Curitiba Museum Oscar Niemeyer, Parana State Authorities from Sanepar, Claspar, APPA, Vietnamese state officials and scientists from MARD, FCRI, PPRI, VEPA and the Hanoi University of Agriculture.

Social Discourse: Full grown installation at ETH Zurich, Switzerland (Photo: Rebecca Naldi)

Window, construction with 294 petri dishes

Sylvia Hostettler

ARTIST IN THE CENTER FOR INTEGRATIVE GENOMICS (CIG) | UNIVERSITY OF LAUSANNE

Disciplines: Sculpture, Installation

Project title: *Light Reactions – Dimensions of Apparent Invisibility.* The aim of the residency was to work on a sculptural installation, which magnified the micro-level of nature and interpreted the behaviour of light on plant growth. Using light boxes and various light sources, works were built to highlight the experiences and observations from the residency. The following issues were important to learn about: The affects of light on the plants, their genetic mutations and manipulations and the analysis methods. The results were shown in the installation in the foyer of CIG itself and generated discussions about genetic topic in the transformation of the artist with the visiting public.

LIGHT REACTIONS – DIMENSIONS OF APPARENT INVISIBILITY Sylvia Hostettler

University of Lausanne

Since 2005 I have been working on a serial project, *Landscapes,* which will on completion be presented in five independent chapters. It tells of unknown places – places where I have spent time and to whose influence I have exposed myself. My application for an artist-in-lab residency, and my placement at the Center for Integrative Genomics, were determined by the research character of the project and the biomorphic forms of my own sculptural work, as well as by its regular references to light. Led by Professor Christian Fankhauser, the CIG team is engaged in fundamental molecular research, in particular with the development, under specific light conditions, of 'Arabidopsis thaliana' (AT: thale cress), a plant widely used as a model in genome research. My project at CIG was the completion of chapter four of *Landscapes,* under the heading *Light Reactions – Dimensions of Apparent Invisibility.*

Phases

My nine month placement at CIG was divided into three phases: Learning: chaos; Conceptual development: tidying up; Realization in the lab and studio.

Learning: chaos (March–June)

The first four months at CIG gave me the opportunity to look into many aspects of the molecular biologists' world and attempt to understand what seemed like a magician's cabinet of complexity. But all doors were open to me and I could move around the building as I pleased. I participated in lab and departmental meetings. Professor Nouria Hernandez, introduced me to the significance of DNA in private tutorials. At first I understood little: it was a language of abstract concepts and abbreviations, an international medium of scientific communication that I was hearing for the first time. I had been given an office of my own with a view over the lake, and I withdrew there to pursue my own thoughts and start my internet research. This only multiplied the questions, however, as concepts unknown to me were explained in a language equally unknown. But I received substantial help from the scientist who had been assigned to look after me, Laure Allenbach (a CIG technician), who explained her own research methods to me and supported me in mine. Thus in this first phase I was introduced to the approach and analytical methods of molecular biology, and in particular to 'Arabidopsis'. What follows is a description of molecular biological methods as I experienced them.

Petri dishes are an important piece of equipment for cultivating cells and bacteria in vitro, or for germinating seeds. In a sterile chamber clean 'Arabidopsis' seeds (either wild type or mutant) are placed in a petri dish half-filled with a sterilized nutrient, and this is then sealed to create a miniature greenhouse. The seeds are packed in tinfoil and allowed to 'hibernate' for a few days in a refrigerated room; then they are moved to an incubator which, although it looks like a refrigerator, is kept at a constant 21°C. Once they have germinated, the seedlings are used for various analyses. For instance, they are taken out of the tinfoil (where they have so far existed in darkness) and subjected to different forms of light. Some proteins, the so-called 'photoreceptors', respond to light signals in different ways, so that red, dark red and blue wavelengths impact and change certain growth conditioning factors in the plants concerned (e.g. phototropism). The early behaviour of mutants

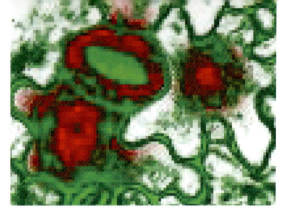
Sylvia Hostettler's draft of the *Window* inspired by stomata

Sylvia Hostettler constructing the *Window* (Photo: Laure Allenbach)

Realization: in the lab and in the studio (September–November)

Like a scientist in her research hypothesis I was immersed in my work on the installation *Light Reactions – Dimensions of Apparent Invisibility*. I spent most of my time on the window, but before the end of November I also had to finish my research in the microscopy department. Four microscopic constellations took shape: *Crossing, Leaf-root, Expression_C8,* and *Expression_C9*. I had devised a medium for the two Expression pieces composed of some strands of my own hair fixed with sticky strips onto a lamella to form a sort of 'tissue' – a word I heard repeatedly over the months of my placement. Selecting a petri dish from the window, I painted its colours onto the tissue and completed the construct under the microscope with AT. I was so pleased with the result that I decided to repeat the process with a symbolic analysis of the window on 294 lamellae that I would set as a light table into a niche in the black box.

Conclusion

I was happy to return to my studio and complete the complex, many-sided project there. Nine months are an ideal time span for acquainting oneself thoroughly with an unknown world and embarking on a major

installation. The final project was shown in the main hall at CIG, and they funded it. The CIG itself, as well as Lausanne's FBM (Faculté de biologie et de médecine), provided generous support. The scientists, too, had enough time to see how in the language of the artist a work of art can grow out of their own familiar material. Although at the presentation in April I could say nothing precise about the form my project would finally take, they could see in the following weeks and months how my office gradually filled up, first with the Internet printouts I pinned to the white walls of the room, then with my own small pieces – singular objects that occasioned a number of conversations. At the beginning of October I showed the current state of the project at one of the weekly 'apéros' in the main hall. I was also present at the retreat in Saas-Fee, where I improvised a 'Postdoc Poster', and at the end of my placement I held an open door day. The CIG people were interested in everything, and I really made friends.

My presence did not change the way molecular biologists go about their work, but it can be seen as a significant widening of horizons. Scientists focus for years on problems of a very small compass; I tried to create something broader and more comprehensive. It was an extremely fruitful education, on which I

Detail, *Expression_C8*, pigment print on paper, 84,5 x 84,5 cm   Detail, *Anomalie 3*, pigment print on paper, 57 x 42,5 cm

shall continue to draw in the future. My experience at CIG will interconnect with new experiences, mutating and surfacing into different forms in the years to come. One concrete idea remains, however: to contact scientists again for a new field research project.

Special thanks to

Laure Allenbach, Christian Fankhauser and his research team, Nouria Hernandez, Nicole Vouilloz, Gilles Boss, Arnaud Paradis. The installation *Light-reaction – Dimensions of apparent invisibility* was made possible by: UNIL/Université de Lausanne; UNIL/Centre Intégratif de Génomique (CIG); UNIL/Faculté de biologie et de médecine; Fondation Leenards; Fondation Fern Moffat/Société Académique Vaudoise; sc/nat, Swiss Academy of Sciences; KulturStadtBern; Erziehungsdirektion des Kantons Bern.

Exterior view of the installation *Light Reactions* at the Center for Integrative Genomics in Lausanne

Installation collage *Autostrada Verde* in the Piazza del Sole, Bellinzona, designed for 2011

Claudia Tolusso

ARTIST IN THE WSL SWISS FEDERAL INSTITUTE FOR FOREST, SNOW AND
LANDSCAPE RESEARCH, BELLINZONA

Disciplines: Scenography, Photography

Project title: *Observation for an Artwork* concerns the landscape and land use – memory and storytelling – nature and architecture. My aim was to research the daily rituals undertaken by scientists in the environment, to document this, and to respond to the experience by creating an interactive installation for the Ticino public.

INSUBRIC ECOSYSTEMS RESEARCH GROUP IN BELLINZONA

Claudia Tolusso

How do our agricultural and economical needs shape our use of land and its biodiversity? What does it mean to speak about 'salvaging nature' in Switzerland? How can we plan our interaction with our environmental resources in the future and how can this be formalized through land management?

The Place

In my first month I shadowed the scientists' experiments and attempted to look through their eyes at the issues and changes taking place in the southern Swiss Alps. I spent nine months in a beautiful three-storey house (the research lab), located on the eastern hill behind Bellinzona. Through the window I had a spectacular view of Locarno and the Ticino river as it cut through the Magadino plain, traced by the forested hills on both sides until it poured into the Lago Maggiore. The researchers at WSL, are spatially and literally embedded in their topics. These are: changes in land use, social economic developments in the chestnut forests, ecological disturbances such as forest fires, biologically invasive ever-green neophytes in the forest, urbanization, and biodiversity in urban areas.

The First Months

From the first day Dr. Marco Conedera, Head of the Research Unit Ecosystem Boundaries at WSL, took me along to his lectures, these alternated between field excursions and the ETH Zurich. At the ETH, he lectured on fire management as well as its ecological repercussions. I also accompanied him and his team to a EU-meeting with an international group of forest fire experts about the need to increase the level of attention addressing forest fires <http://www.

wfc2009.org>. There, I discovered that the problems wrought by climate change are increasing the need for solid fire management. It was mentioned that some increase in the frequency of forest fires can indeed be attributable to climate change. Other issues targeted the direct (regional) neighbours of Switzerland, as outlined in the dissemination plan in Milan <http://www.euro-fire.eu>.

Through my visits to various conferences I became interested in questions like: How can costs be secured to allow the continuation of this research; or how subjective (objective) can a scientist be when he or she is involved in such important topics; or how vital is it to disseminate highly specialized research or be subject to public opinion? Often, this last question is a political issue, as the capacity for scientific research to effect any feedback at a social level could colour their findings and, eventually, affect the level of funding for such projects in the future. It seemed to me that anything I was going to propose should somehow raise the awareness about the importance of this funding.

Methods for Scientists and Artists

I made a small studio for my work in the library and started the process of developing my own strategies and creative ideas that could communicate these topics. I also delved more deeply into the methods of scientific observation, taking a special interest in the methodology that was used by the scientists present. I became very fascinated by a procedure called dendrochronology, used to calculate the age of the forest and how the weather patterns have

Excursion with the Insubric Ecosystems Research Group in a burnt area of the Ticino forest

Excursion on Biologically invasive evergreen neophytes in the forest of Ticino

changed over time, as well as other information. I was also introduced to palynology: a way to track carbon particles in lake sediments, which clearly indicate the evolution of ecosystems and fire regimes. As an outsider I started to understand that the natural ecosystem is extraordinarily complex. The scientist's use of technology permits a look into the past in order to understand the future, its development, and perhaps even allows him/her to shape it. But what did it mean for the experts to see the past through such views of the present? What problems are now on the hot list and how complex is it to try to solve them?

I began to see that my artistic tendencies could address some of these questions and engage an audience at different levels. During my involvement in the project, I read the book: *Into the Next Dimension*, by Clemens Kuby, in which the Heisenberg uncertainty principal is discussed. ('Dass der Betrachter durch das Betrachten das Betrachtete beeinflusst') (He stated that the viewers presence can also influence the object of his observation.) Then I came across another phrase, 'Absence of evidence does not constitute evidence of absence.' These quotes led me to make a sketch of an idea for a marble work on the flat rooftop of the WSL Institute itself.

Over the next few months, I decided to record and observe how the scientists work by using a camera and a sketchbook, a process which furthered the relationship to my immediate environment. These photos captured many impressions about the scientific research and I continued to read numerous papers and publications. Slowly, the scientific terms and their meanings became easier to understand. For example, one of the words Marco Conedera often used was the verb 'to model' and the use of it in the phrase – 'the modelling approach', by which he meant – one has to collect and analyse data in order to demonstrate or even prove ones thesis.

I learnt that data sets led to discussions about their exchange within the Institute, as well as with others beyond the borders of Switzerland. With another scientist from the EU fire projects, Gianni Boris Pezzatti, I had rather philosophical conversations about the processes within their research projects and how nature has its own taxonomy.

In May 2009, the Research Unit Ecosystem Boundaries met to exchange their results in Bellinzona. At this meeting I had the opportunity to present my ideas to a broader group of scientists, who came from other WSL institutes (Davos and Lausanne).

Claudia Tolusso in the field observing the scientists working in their environment

Work ideas in the artist's studio at the WSL in Bellinzona

This was an entirely new experience for me, and the reactions from the audience were surprising. I showed them sets of photos that I had taken of them in their working environment, as well as some sketches and models that they used, and took a rather metaphorical approach to their research topics. Along with my images I used the following quotes:

- 'Nature manifests itself when it rains, culture manifests itself when one has an umbrella.' Franz Josef Wetz
- 'Only we are accountable to the laws that we invent to describe nature, not nature itself.' Prof. M. Sieber, Lehrwald (ETH Zurich)
- 'The map is not the land.' Gregory Bateson

Their comments indicated that my work was a poetical eye-opener for their topics. Patrik Krebs, from WSL in Bellinzona, remarked, 'You are free to play in a visual and philosophical way but our output has to prove our thesis!' Dominik Kulakowski, Assistant Professor (Clark University, Worcester), commented that my earlier photos from the forest in Melbourne manifested an unfamiliar viewpoint about the realities of nature. In fact, we have maintained contact and still discuss the correlations between science and art. By contrast, the scientists' presentations were focused and defined by a certain question, and the

data sets were collected and analysed to answer that question. It was fascinating to be able to follow their exchanges, and it seemed that their presentations and gatherings were a very effective way to keep up to date.

We also had encounters with the other artists-in-labs-residents; this is where I came across Jorge Luis Borges' book *The Map*. These exchanges were important, as they allowed us to share doubts and/or questions about the different processes. Beyond that, I shared a flat nine months long with Alina Mnatsakanian, a resident of the AI lab in Lugano, and we became good friends. I am very happy that I received feedback from the artists about my images and about the problems of interpreting rather than illustrating scientific research. For me, photography is a 'wordless discovery' about the life of a scientist. I worked to communicate this content and make an artistic and interpretative response.

The working title for my final project was *Observations for the Realization of an Artwork*. In it I proposed three installations to the city government of Bellinzona for the *Vivere i Sensi* art event in 2011. In January, 2010, Marco Conedera and I went to discuss further details with the curator Flavia Marone,

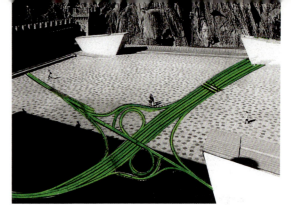

Installation collage *Seed Explosion* at the Castello Grande, Bellinzona, 2011

Installation collage *Autostrada Verde* in the Piazza del Sole, Bellinzona, 2011

responsible for culture in the city of Bellinzona. All three installations should take place during April 2011, the date of the twentieth anniversary of the WSL institute in Bellinzona. They will be part of a series from the Federal WSL Research Group.

The first Installation (*Seed Explosion*) targets the problem of biologically invasive evergreen neophytes in Ticino. It will be located in the Courtyard of the Castello Grande. A two-meter weather balloon will be released on a twenty meter tether and explode, showering rice paper seeds onto the ground. This work addresses the questions: what is an invasion? And what is the process of phytogrowth? For me, this place and the balloon creates a metaphor showing how uncontrollable an invasive species really can be. The Castello Grande was originally built to protect and defend Bellinzona from invaders!

In the second installation (*Autostrada Verde*), the subject of biodiversity in urban areas will be addressed. This will take place at the Piazza del Sole in Bellinzona. On the 2500 square meter piazza, I intend to install a grass (not concrete!) relief in the shape and form of the South Cross Highway (scale of 1: 300). This piazza is one of the most frequented non-vehicular crossing points, and also a place of

encounter for skaters, bicyclers and daydreamers. However, *Autostrada Verde* inverts the speed of a normal highway, because it allows the audience to slowly contemplate the spatial and visual segregation of its landscape and creates a social space where diversity can flourish. The movement and use in this space will be documented.

The final installation *Data Cloud* is based on the idea of how research data could be visualised. It will take place at the Bellinzona City Hall, in the courtyard of a three-storey building, constructed in 1680. I will project a cloud onto a set of umbrellas. During the day this object will project an image of the real clouds as they pass over the quadrangular and roofless courtyard; at night the projection would display the process of cartography and the single data sets that make the process readable. The project demonstrates the transient nature of data.

Conclusion

I am particularly excited about the realisation of the Bellinzona event, and am looking forward to an interesting and challenging collaboration with the local public, as well as the unpredictable moments. I experienced a field of science to which the general public has very little access, and whose discoveries are only

Model photo of the Installation: *Data Cloud* in the 'municipal' of Bellinzona, 2011

Model photo *Data Cloud*, 2011

known to a chosen few. I was surprised how much manpower scientific research requires, and it gave me great satisfaction to carry out research on my own time within such an environment. As a result of this experience, my artistic horizons have expanded, and my interest to collaborate with other institutions has grown and I have become aware of the value of my art work.

Thanks to

The WSL employees for their openness and warm welcome from the beginning. Special thanks to: Daniela Furrer, Cinzia Pradella Janet Maringer, Begona Fijeco, Damiano Torriani, Boris Pezzatti, Eric Gehring, Patrik Krebs, Marco Conedera and Marco Moretti.

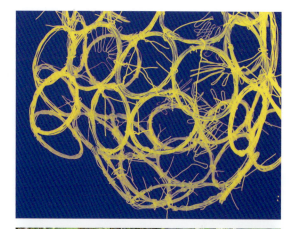

Art object inspired by the construction of the rock-fall-grid in the forest of Ticino

Watercircle Toilet Fountain inspired by Eawag's no-mix toilet (Photo: Pierre-Henri Dodane)

Ping Qiu

ARTIST IN THE EAWAG: THE SWISS FEDERAL INSTITUTE OF AQUATIC SCIENCE AND
TECHNOLOGY, DÜBENDORF

Disciplines: Sculpture, Land Art

Project title: *Bathing.* My aims were to learn about water and ecosystems in the developing world, and immersion in the science lab, then use this inspiration to construct a set of two installations and one performance. I wanted to build a public fountain about the eco-potentials of bath-water in different countries, and to trace how humans can interfere with aquatic systems.

BATHING Ping Qiu

March to June

My name Ping means 'a little green thing, which lives on the water'. At first I thought that I was a kind of algae, but soon the scientists told me that I am a little bit bigger than any algae they have ever seen. For my residency, I was offered to learn about three levels of scientific research and collaborate with the scientists involved. These were Dr. Christopher Robinson from the ECO Aquatic Ecology group, Dr. Christian Zurbrügg from Sandec, Sanitation in Developing Countries and Renata Behra (UTOX, Environmental Toxicology). These three scientists also cooperated closely with one another to provide a lot of valuable information for me. I soon decided that the three labs were like small screws in one big machine, all contributing to the same overall objective: the understanding of Aquatic Science and the potentials of Technology to help solve ecological problems that they have found. Because they share results from each other's research, I wanted to develop a complex installation based on the inspiring information from all three labs. Although this may have been a nice idea it was not easy to work with three labs at the same time. It was so much information – like a flood – that soon I felt that I was getting too drunk. I visited many different lectures at the beginning but I knew that if I went to every lecture it would be like shooting without aim, and I would lose myself if I did not make some choices or get some help. So during the first months, Chris (Dr. Christopher Robinson) kindly arranged for Debra Finn, a post-doctoral student at Eawag, to give me solo lessons in Aquatic Ecology. In addition, Tati (Renata Behra) gave me lessons about nanoparticles. This soon helped me to understand some of the scientific research that I experienced around me. Therefore, I tried to learn as much as I could. During these months, I also worked in the lab of Environmental Toxicology and Aquatic Ecology with the light microscope and became fascinated by this world and I also attended more than 15 lectures by different Environmental scientists.

July

In July Dr. Christian Zurbrügg sent me to Dakar, Senegal with Pierre-Henri Dodane, the scientist at Sandec, to see how they work with public sanitation. This was an important learning experience about Clean Water Programs and I documented my experience on video and conducted interviews with people there, which I later made into a documentary film. Two days after I returned from Africa, I went with a research group of Aquatic Ecology to Macun in the Swiss National Park at the top of the Alps, and witnessed how hard scientists also work in this field. Before I met the scientists at Eawag, I had a preconceived image about how scientists might look and think, wear glasses and do calculations. I thought they would not get along well in society and would not face the problems of daily life. The trip that I took to the Alps with Chris changed my idea about scientists. I noticed that he was very practical in the mountain environment and that he could easily survive in the wilderness. On this trip, he took very good care of me – even supplying mountain boots, rain parka, and sleeping bag, knowing that I had no experience with this level of survival. In this way I have not only learnt about biology in my residency, but I have also learnt many lessons for everyday life. From these field work experiences in the mountains and in Senegal, three ideas for new artworks emerged:

Dragonfly larvae: Odonata, and a Hymenoptera: stereo-micro-scope

Ping Qiu in Dakar, Senegal (Photo: Pierre-Henri Dodane)

nutritive organic molecules from inorganic sources via photosynthesis. This can involve light energy (hence my egg sculpture). They saw me as a self-sustaining organism because I seemed to have the ability to synthesize my own food from 'inorganic materials'. The EMPA labs that had all these materials was across the road from Eawag. Also they saw me as a primary producer, because I attempted to introduce new organic materials into their environment that the primary consumers (scientists) could feed upon!

Overall, the scientists gave me a great amount of their energy and I was deeply inspired by their environment. My art production certainly flourished from this energy. Many scientists at Eawag also expressed an interest to learn something from artists and so I formed a group called *Science meets Art* including artists-in-labs co-director Jill Scott, to plan these potentials for the future. We may work with different images, but many working processes and attitudes towards creativity could be very similar.

Credits

Christopher Robinson (ECO Aquatic Ecology Group), Christian Zurbrügg (Sandec, Sanitation in Developing Countries) and Renata Behra (UTOX, Environmental Toxicology). For special help: Herbert Güttinger (Direktionsstab Eawag); Sandra Ziegler Handschin und Andri Bryner (Kommunikation, Eawag); Daniel Pellanda, Raoul Schaffner and Canan Aglamaz (Informatik, Eawag); Peter Gäumann and Andi Raffainer (TD Werkstatt, Eawag); Christiane Rapin Nussbaumer, Vicenç Acuña, Mark Gessner und Tom Gonser (Eco, Eawag), Ilona Szivak, (UTOX, Environmental Toxicology); Erich Eschmann and Atila Redondo (Logistik und Infrastruktur, EMPA); Jack Eugster and Kai Udert (Engineering, Eawag); Pierre-Henri Dodane (Sandec, Eawag); René Crettol, Joep Breuer and Rolf Luchsinger (Center for Synergetic Structures, EMPA).

Watercircle Toilet Fountain (Photo: Andri Bryner)　　　　*Watercircle Toilet Fountain*

Eggs' Breath installation: Ping Qiu and Christopher Robinson in the workshop in EMPA, near Eawag

Interactive sound installation *Kinetism* at the Faculty of Architecture of EPFL, 2009

Luca Forcucci

ARTIST IN THE BRAIN MIND INSTITUTE (BMI) | EPFL, LAUSANNE

Disciplines: Sound Art, Composition, Installation

Project title: *Music for Brain Waves:* The essence and inspiration for this project came from a work called *Solo Performer* by the American composer Alvin Lucier. I was interested to extend this research by investigating the addition of 'loop feedback' as a spatial component and to extend parallels between art/science from the 60s with today's software and EEG (electroencephalography) hardware technology.

MUSIC FOR BRAIN WAVES Luca Forcucci

March–April

I spent the first month immersed in my new environment and I was surprised to discover that there were 3 labs who were intensely interested in my presence: The Laboratory of Neuroenergetics and Cellular Dynamics (LNDC), The Digital Holographic Microscope group (DHM) and the Laboratory of Cognitive Neuroscience (LNCO). I made a tour and as an initial introduction, the scientists from these labs explained the aims and methods of their respective research foci. The researchers at the LNDC gave me an introduction to the terminology and the composition of parts of the brain and their respective functions. One of their main achievements of research is the discovery of the astrocyte-neuron lactate shuttle by Prof. Dr. Pierre Magistretti. Pascal Jourdain from the DHM lab taught me about holography and how the microscope works to visualize cognitive stimulation. The team at LNCO introduced me to their research interests, which are based on the body and the potentials for self-consciousness. The vocabulary of such research is vast and I had to attend seminars, read papers and books in order to understand what they were talking about.

I was taught to record EEG signals (electroencephalographs) through a software program called *Matlab* and a hardware device called *Bio Semi*. I discovered and investigated an important brain area for my own research on emotion: the Amygdala. My understanding of this area was further informed by talks with post doc researchers, papers and books. Finally, I was invited to attend a classical concert of Sibelius by Estelle Palluel, one of the post doc science students from LNCO, who is also a classically trained musician.

That musical event was a nice omen and I started to feel at home. For my main project and composition, I investigated the relation between auditory perception and brain waves activity. What kind of experiments would help me study this reaction according to the nature of the sounds heard?

Though I kept in mind my initial proposal *Music for Brain Waves,* I decided to investigate other relations between the main researches of the LNCO lab and sound art. After a few interesting discussions with Olaf Blanke, who is the head of the LNCO lab, he informed me about the state of their research and provided me with papers and books to read. I conceived of a project that would be an investigation between an 'out-of-body experience' and sound. Since the beginning of the residency I met regularly with 2 PhD candidates from the LNCO (Pär Halje and Oliver Kannape) for specific knowledge about neuroscience and by April I investigated mental imagery, emotions and EEG as main subjects. I was introduced to mental rotations and perceptual anticipation. I attended a seminar on neuroscience given by members of the LNCO (Olaf Blanke, Jane Aspell and Christophe Lopez) during one week with other PhD students in cognitive neuroscience. There I learnt more about brain neglect, various pathologies, self consciousness, body illusions and the multisensory self studied by the laboratory of cognitive neuroscience. I was becoming fully immersed in the scientific community. Mimesis was happening with them because of my full immersion in the lab and I had the strange impression that I lost my art cues but surely this was an illusion!

During a talk on space and neuroscience given by

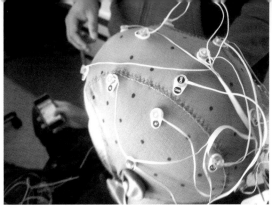

EEG recordings for an experiment with sound listening

Live spatialization of sound (Copyright: Pierre Pfiffner)

Olaf Blanke at the faculty of architecture, I was introduced to architect Isabella Pasqualini, who was also conducting research in neuroscience. This meeting evolved into a great artistic exchange and we shared many issues about the perception of space. Her lab *Alice* is located at the faculty of architecture and she invited me to give a lecture. I presented my sound work and artistic research related to space as well as my inspirations from the residency at the LNCO.

May – July

I went back to the DHM to learn more about their research and how I could possibly relate it to my project. I asked for several videos from the holographic microscope as material for my compositions. I also learnt the vocabulary related to the death of a cell (neuron), how to stimulate cells and some biophysics aspect analyses about them. I decided that I could possibly transfer their images of stimulated neurons into sound. For the movement of the neurons would in fact create its own composition, and the sounds produced could then be manipulated by the brain waves by using the EEG. Therefore using *Max/MSP,* I started to code an instrument that was able to do that manipulation. From then on my neuroscience meetings were split between information on cognitive research and specific EEG recording

aspects. During the cognitive approach of neuroscience, we discussed the models for multi-sensory perception and integration as this really helped me to investigate the function and anticipation of sound tasks and I also attended more lectures about brain pathology. After 3 months, I started to really understand the research of the lab, and more ideas came into my mind.

Together with a member of the LNCO (Tej Tadi), we were invited by the architect (Isabella Pasqualini) to participate in a project involving neuroscience, architecture and sound art. The project would cross the boundaries of the labs by working on a multi-disciplinary approach. The art gallery at the faculty of architecture asked Olaf Blanke and I to present a collaborative work of art. We responded with a sound installation based on the idea of the auditory perception where internal and external spaces could be audible at the same time. How could we explore normal inaudible or neglected spaces and sounds? The exhibition idea resulted in a further research project that would attempt to cross the boundaries of architecture, neuroscience and sound from a transdisciplinary approach. At this point a Japanese scientist called Sachiko Hirosue contacted me, who was researching at the BMI and is involved in a festi-

Holographic microscope manipulations (Copyright: Laboratory of Professor Magistretti, Brain Mind Institute, EPFL)

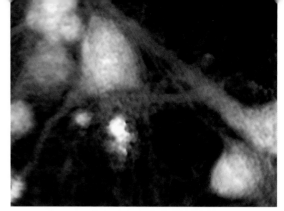

Images of neurons from the holographic microscope (Copyright: Laboratory of Professor Magistretti, Brain Mind Institute, EPFL)

val in Toronto called *Subtle Technology*. We started a dialogue about art and science. Also, I decided that if I wanted to continue my research after my residency, and also perform with the EEG interface, that I may have to buy my own more portable and wireless EEG system.

The project based on architecture evolved into a site-specific project exploring the perception of time versus the time of reaction, involving urban and architectural issues, light perception and sound art. I investigated philosophical questions by reading Deleuze's and Guattari's *Mille Plateaux*. I worked on my sound installation by making field recordings (body and urban sounds) during my spare time. I continued also my lessons to understand mental imagery.

August–September

In these months I found myself working on a few sound art projects simultaneously: The first one was *Music for Brain Waves*. The interface and code was the software for compositional purposes. The second project was *Kinetism* an installation, which explored the ways of composing sounds of the urban environment with body sounds and the third project was MSEC an audio visual haptic installation which could merge architecture, neuroscience and sound

art. Here the prototype was designed for an urban bridge. However, *Kinetism* the collaboration with Olaf Blanke was launched at the exhibition opening and it was the first artistic achievement of the residency. We attempted to achieve an 'out-of-body experience' through sound, and according to some comments by the viewers they had this experience.

During September a new computer scientist and virtual reality specialist called Bruno Herbelin came to the lab and he had worked with artists. We quickly connected. He showed me a haptic device under development with a European project (*SUM, Structured Understanding of Music,* a Scandinavian research project leaded by L. Graugaard) and they were looking for a composer, so we started to discuss possible collaborations. The project may result in another interactive audiovisual installation. In relation to my initial project *Music for Brain Waves* my goal became to understand which parts of the brain were activated, when a sound was proposed to a participant. This experiment could essentially help me to compose new sound compositions. I designed the experiment with a member of the lab, who helped me to code *Matlab* software in order to trigger sounds and record EEG responses. I made a psychological questionnaire with another member of the lab so that

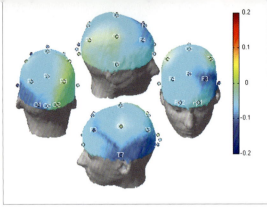

Results from my experiments: EEG activation by sounds

I could analyse the results on those who undertook my tests. I collected sound and images from the fMRI scanner during the last week of my residency as raw material to re-use in other compositions.

Conclusion on Art and Science
In this residency, I discovered that art and science share a main tool called 'observation'. The only difference is that scientists split observations into many slices and analyse each one, while artists tend to do the reverse – we start with many slices and work intuitively backwards to try to make one observation.

I had the opportunity to invent 5 original projects, inspired by the knowledge I gained. I am glad that I let this inspiration happen and that I deviated from my original proposal. From my point of view, the residency resulted in such a full immersion in neuroscience, that I now feel over stimulated. In the next months I have to digest the information and knowledge and although one work has already been shown I will present more results in the near future.

The value of putting artists into scientific research labs must be observed as a long term effect and this effect may not even reveal itself until 5 or 10 years have passed. Finally in the last weeks of the residency,

Olaf Blanke, Sebastien Dieguez and myself wrote an article together and published it in Nature – called *Don't forget the artists when studying perception of art* (Nature, vol. 462, no. 7276 (24 December 2009), p. 984).

Thanks to
Pierre Magistretti (Director of the BMI), Olaf Blanke (Laboratory of Cognitive Neuroscience, EPFL), the team from LNCO, Pascal Jourdain (Digital Holographic Microscope, EPFL), ALICE – Atelier de la conception de l'espace (Section d'Architecture, EPFL), Dieter Dietz (Alice, EPFL), Isabella Pasqualini (Alice, EPFL), Jill Scott and Irène Hediger.

Kinetism, interactive sound installation, 2009 (video still)

Luca Forcucci listens to the dual sound tracks of *Kinetism,* 2009 (video still)

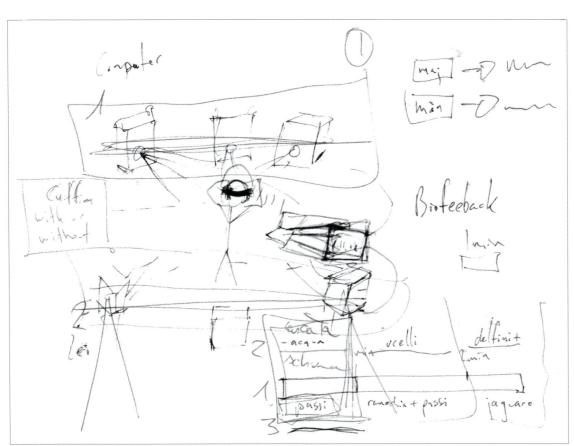

Sketch and ideas for the *Music for Brain Waves* work

@Location

PAXman helps Constant Traveller to justify if his travel is unavoidable and it helps him to optimize his travel itinary.

PAXman`s Agents

location sensors assist, respond to needs and guide PAXman through networked social space in the airport.

@Airport

PAXman 2.0 suggests many social activities depnding on Constant Traveller`s mood, time available and personal interests.

Travel Itinary

Mood, Time, Interests

Travel Optimizer
- arrange meetings with other PAX
- pre-plan airport activities
- offer social activity at airport
- book social activity at aitport
- arrange plane seating with PAX of simmilar interests

Social Activities
- „slow" communication with other PAX
- ad-hoc meeting witn other PAX
- co-working and workshops
- clubs of interests
- fitness activities

my Travel Impact
- my Heallth/Fitness
- my FoodPlanner
- my Emotional Experience
- my Earth

my Barometer
- my PAXman scores
- my eco-performance
- my PAXman community

Augumented Airport
- emotional map of other PAX
- overview of all current activities in Constant Traveller`s proximity
- airport „ open data"

PAXman 2.0 – an idea for a context aware mobile application for 'constant travellers' in the airport

Monika Codourey

ARTIST IN THE HUMAN COMPUTER INTERACTION LAB (HCI LAB) | INSTITUTE OF PSYCHOLOGY | UNIVERSITY OF BASEL

Disciplines: Architecture, Media Art

Project title: *Constant Travellers*. The aim of the project was to develop a prototype for a pervasive game about the effect of mobility on our emotional states. The source of inspiration for the design was to study the impact of physical and informational mobility of 'constant travellers' on their perception and behaviour while they were travelling for business purposes.

CONSTANT TRAVELLERS Monika Codourey

Relevant Questions

The lifestyle of constantly travelling business subjects is increasingly dependent upon transit spaces woven together as spaces for work using laptops or mobile office phones. As a result many economic and social aspects of business travellers' everyday life are increasingly conducted 'on the move' or away from 'home' and in the 'space of flow'. In addition this mode of de-territorialization involves a progressively greater population and results in emergence of new socio-technical practices (e.g. surveillance, social sorting, biometric authentication) influencing patterns of movements and co-presence within transportation systems. Transportation corridors are often considered as 'non-place' of transitory nature, empty of any meaning and social significance. Studies show that geolocated technologies augment the urban environment of their users with informational resources and therefore they are excellent tools to reshape the experience of mobility in urban settings. Before I entered the residency I had many questions. Can the mediation of handheld devices convert the transitory nature of constant mobility into a creative experience itself? What kinds of interaction supported by mobile technologies is most suitable for development of game content based on the geolocation principle for 'constant travellers'?

What are the movement and interaction patterns of business travellers? How do they utilize mobile devices in everyday business and leisure activities on the move? How do business travellers perceive the space of flow during their everyday activities? What kind of social and spatial relations do they experience? How fast do they go through security and safety thresholds? What kind of emotions/reactions are connected to border-crossing procedures and biometric authentication? What impact do new socio-technical practices have on their life?

At the Human Computer Interaction Lab (HCI Lab)/ Institute of Psychology at the University of Basel I wanted to know how psychologists successfully formulate questions for interviews. It also seemed to be a great opportunity to learn about the scientific methods applied in Psychology and human computer interaction (HCI). I wanted to know how I could apply these methods to my own working process.

March 2008

I remember when I began my residency in the Human Computer Interaction Lab (HCI) on March 4th in 2008 and was introduced to the HCI team, consisting of five members who mostly worked together in a small office space. Sandra Roth and Alex Tuch were my scientific advisors and they gave me a tour around the Institute for Psychology facilities and showed me the Usability Lab for their research experiments. The first weeks were filled with learning about their psychological approach to HCI and their usability tests, research projects and working methods. It was immediately clear to me that the word 'experiment' is a very loaded word in Psychology. Moreover, Peter Schmutz integrated me into his weekly seminar and included my project as a case study for evaluation of appropriate research methods for his own students. It turned out that my investigation should consist of multiple studies. At first I should conduct a series of qualitative interviews with ten to fifteen people who I consider as 'constant travellers'. These

Data Record of Mobile Identities <http://mobile-identities.info> interactive installation for quantitative survey on 'constant travellers'

Monika Codourey during an interview (video still)

results would be a starting point for the formulation of my thesis about what their needs are in relation to mobile interfaces. However, to support my findings, it would be necessary to conduct quantitative studies with over a hundred people.

April 2008

In the second month of my residency, I studied how psychologists conduct their research, I read their books and asked endless questions during regular weekly sessions. Moreover, I conducted an extensive literature research on mobility paradigms and classification models. Consequently, I formulated selection criteria and a list of open questions for the qualitative interviews with the 'constant travellers'. These questions were grouped by the following topics: professional background, their travel pattern, mobile lifestyle, emotions, and behaviour in transportation hubs (airports and railway), and personality. The next step was to find potential interviewees according to defined selection criteria and to arrange interviews with them.

May 2008

In May, I managed to conduct my first interviews with 'constant travellers'. I was very excited about the whole interview process and I found listening to

people's stories very inspiring. My scientific advisors were also very excited about the fact that I managed to engage people to participate in my research so easily. Apparently, this is one of the major concerns for psychological surveyors.

After word-for-word transcribing of the interviews, my advisors introduced me to a qualitative research tool called *TAMS Analyser* or the *Text Analysis Mark-up System*. This software is specifically designed for use in ethnographic and discourse research and it favours issues of objectivity in science. As the psychologists suggested in order to avoid any bias, different people should analyse the collected data. For me, the most important aspect was to comprehend the meaning of the scientific language, one that actually ignores claims or statements and instead describes findings as possible indicators. These requirements may be too ambitious for one person, but it was interesting to borrow some of the scientific methods for my research.

However, I realized that it is not necessary to follow all scientific conventions as a media artist and architect. As a next step, I took the careful design of questionnaires from science, as this was necessary to support my findings by using quantitative research.

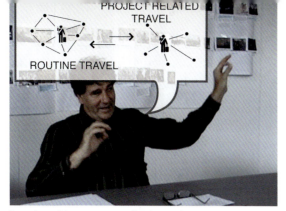

Interview with constant traveller 1: Travel Space-Time

Interview with constant traveller 2: Work Space

June – July – August 2008

In the months June and July I conducted and evaluated more interviews by using the mediums of video, e-mail and Skype, but I found that talking to people 'face to face', was by far the most informative method. However, e-mail interviews were the easiest to evaluate. In these ways several interviews with heterogeneous groups of people who travel extensively as an integrated part of their work, were conducted. The interviewees came from every age group and were gender balanced, aspects, which I learnt, is an essential aspect for a complete qualitative survey. After the first raw cut of video material I could draw first conclusions. Next, I had to learn how to design quantitative surveys and I decided to use a conference and exhibition in Singapore (Lucid Fields) which was organized by the artists-in-labs program in order to collect these responses. I learnt from the psychologists that the formulation of questions for quantitative studies is a very difficult process, especially if one wants valuable results in statistical analysis. Consequently, they taught me how to redesign my questions and I have a new perspective from this experience. The results of emotional, cognitive and behavioural states of 75 travellers were collected over 5 days and can be seen on the data record of Mobile Identities <http://www. mobile-identities.info>. This survey also proved that the 'constant traveller' label could be defined.

September – October – November

I continued to conduct quantitative surveys by using the Institute for Psychology local databank of Swiss volunteers. As a result another 60 people participated in the survey over 1.5 days. Therefore, research data from 2 different contexts was collected and compared: an art exhibition/conference and an online databank of volunteers. It seemed that the exhibition/conference provided more variety of constant travellers, while the volunteers from the psychology databank yielded a large quantity of results in a very short time, but less variety of profiles.

At the beginning of October, the scientists showed me how to do statistical analysis of collected data using *SPSS* statistics software. Psychologists can even focus on the study of statistics for their PhD research, but when I fed my statistics into this system I was happy to be confirmed that 'constant travellers' existed! My advisors even expressed an interest in writing a paper for a scientific journal next year about these results. After this I turned back to the original prototype idea I had for a pervasive game-like project for the 'constant travellers'. Unfortu-

PAXman 2.0 Airport Tagging Game: Slow communication/intelligent Furniture

Augmented airport of the travellers' emotional experience

nately, I could not get much support from the HCI Lab in this regard as for them mobile computing is only a part of their usability studies and they do not actually develop interfaces. From my previous research into pervasive game design, this study was a very essential component of human-centred design. Nevertheless, I wanted my prototype to also strongly reflect the user-centred research approach from the HCI Lab. The HCI Lab focuses on usability studies of web-based interfaces with support of eye tracking software. So I used the opportunity to test my mobile-identities interface. I was impressed with how just a few usability tests could help to improve a design and make it more efficient.

Conclusion

My residency not only taught me about the methods used in applied psychology it also showed me that working together with scientists was an intense exchange and an extremely inspiring process. Being nurtured through this process enriched my way of working and gave me more confidence. The experience has lead to development of two projects. A formal video essay from the interviews and the design of PAXman 2.0 – an artistic prototype for the public space of the airport itself. This is aimed at changing the presumed behavioural patterns of 'constant travellers'. The intention of the project is to implement pervasive game elements in order to localize the use of mobile social media in the physical space. The artists-in-labs residency was a great experience. And this led to an award of another Residency in the UK with a leading interactive media research group called Blast Theory. In the end it was difficult for me to leave because I was treated like a kind of guest scientist. Certainly 9 months was enough time to get a general idea about the HCI Lab's cognitive psychology research and make interesting contacts for the future. I was truly impressed how quickly design mistakes can be detected to improve the usability of the interface.

Credits

Sandra Roth, Alex Tuch, the HCI lab, all the 'constant travellers' and all the travellers who answered online questionnaires, Jill Scott and Irène Hediger and the BAK for financial support.

Ad-hoc meeting with another *PAXman* user

Co-working at airport through *PAXman* users

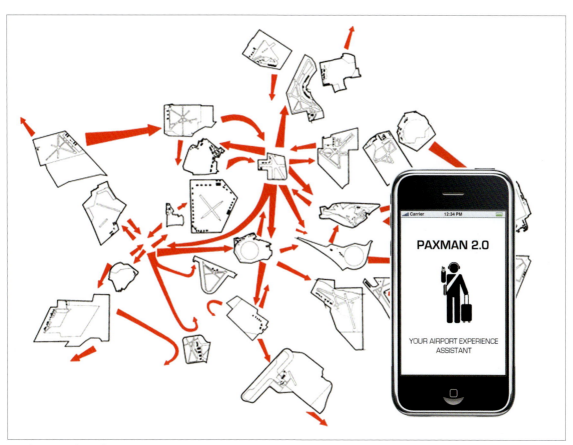

Initial Design Scheme for *PAXman 2.0*

QUARC – installation in the foyer of the Physics Department at the University of Geneva, March 2010

Christian Gonzenbach

ARTIST IN THE PHYSICS DEPARTMENT AT THE UNIVERSITY OF GENEVA | CERN

Disciplines: Conceptual Art, Sculpture

Project title: *I wish I was a physicist.* My residency involved finding an innovative way of representing the non-visible. My research in the lab would also be based on the scientific perception of non-visible matter. How can we understand what is not imaginable, how can we represent the immaterial substances, how can we make art about antimatter?

I WISH I WAS A PHYSICIST Christian Gonzenbach

The naïve start

I made my own way through physics, by trial and error, to come to terms with my presence as an artist in a scientific environment. I imagined that entering this physics laboratory would be like jumping into a swimming pool. But after I came to the surface for the first time, I discovered myself in the middle of an enormous ocean, driven by the streams of all different fields of physics. Swimming in the ocean was exciting so I observed all the currents that took hold of me: extra-dimensions in cosmology and string theory; dark matter that holds clusters of galaxies together; the structure of all matter in small and large scale; the four known interactions; quantum physics and the uncertainty principle; topologies and manifolds; black holes; voids; vacuum energy. I actually had become a student without a class – a lucky freelance student. I was already reading theories by Brian Green, Roger Penrose, Einstein, John Barrow, Stephen Hawkins and I was able to discuss my readings with the physics professors in the lab. I enjoyed this new ocean, happy because I could swim and not drown in information; although, I was a bit concerned about what I could possibly do with all that knowledge. A residency for an artist is often a place where he/she has access to facilities that he/she is not used to, like an etching press or a steel foundry. But I had access not to facilities or skills, but instead to knowledge – the deepest knowledge of our universe. What can an artist do with knowledge? Is knowledge a raw material that can be shaped like steel or manipulated like the real things I usually make as an artist? I started my residency with a naïve project: *Sculpting Dark Matter*. I had always dreamt of sculpting dark matter and hoped to make my dream come true with the help of the physicists. However I soon learnt that dark matter might not exist. Of course, there is a lack of gravitation in the universe but there is still no evidence of dark matter. For me, this meant that I could not sculpt it: as long as it did not exist in the lab and therefore, I could not use it for my artwork. I was lost. What could I possibly use instead? Particles or bosons, fermions or hadrons? I soon learnt that I couldn't use these either. While no sculptor can avoid using particles because everything is made out of them, to sculpt them was not possible either, because well, nobody can actually see them with the naked eye.

Therefore I wanted to find something that contemporary physics was not able to deal with? My aim became to somehow connect my artistic practice with contemporary physics, especially those borders where physicists reach the limits of their own practice. I was not interested in creating images of what was known but instead look for the unknown, the 'terra incognita'. I quickly realized how far this 'terra incognita' was from my own knowledge and I had to accept the fact that I could not become a physicist in just a few months. While today's physicists already try exceedingly hard to reach the unreachable, they keep looking and recording in order to analyse every possible thing and everything perceptible by using traditional observation tools like highly sophisticated microscopes, spectrometers, voltmeters and many other multiple-measure-meters. New, more powerful tools have also been invented like the great LHC at CERN. With Professor Martin Pohl, I was even able to follow the construction of The Alpha Magnetic Spectrometer 2 (AMS 2) at CERN: a particle detec-

The artist in the 'AMS 2' physics lab

Tracing the trajectories of the particles

tor for the international space station. The device is able to collect and measure outer space particles in a similar way that Vladimir Nabokov once collected and categorized butterflies in the 1880s. The construction of such a machine is amazing and complex and many researchers work for many years using only the most advanced materials and skills. I learnt that physicists really are tool builders and they make tools to analyse the world. Artists also analyse the world, but while physicists classify the world into numbers, artists shift this analysis into images. If I wanted to explore the unknown I needed my own apparatus, my own synchrotron, but what unexpected unknown was I looking for and where could I find it?

Jealous of the physicists

I soon became jealous of the physicists, because they not only worked on the best topics: they also looked for the real secrets of our world. They collected real matter from outer space, plus they could build the biggest devices humans have ever built on and off the planet. Art seemed to be a poor case compared to physics. I suddenly felt powerless in front of them and paralysed. By May I had studied cosmology and extra-dimensions with Prof. Ruth Durrer; nuclear physics, vacuum theory, anti-matter and the relation of space and time with Prof. Martin Pohl, dynamic

systems with Prof. Eckmann and Prof. Wittwer and quantum teleportation with Olivier Landry. I played with all the incredible experiments from the Physi-Scope with Dr. Olivier Gaumer. More importantly, beyond the raw learning of physics, I met wonderful people and shared their passion for fundamental research. People often asked me if I was inspired by physics and I answered: 'I am excited rather than inspired: because particles are not inspired by each other, they are excited.' I told them I felt that I shared their deep curiosity for the world and its components. Still, I could not get rid of a strange feeling: the physicists around me knew secrets that I did not know about how the world actually works. Then one day Martin Pohl told me about the existence of the void inside all matter, the existence of so much space around each particle and explained how matter holds together because particles are constantly exchanging their photons. If we could visualize the scale of particles we would see the photons as fireworks bursting into the empty space. All things exist only through their interactions.

Back to my own language

At this point I realized that I had to go back to my own language in order to talk about physics. Artists always search for new ways to look at our world and I am not

Matter made out of space

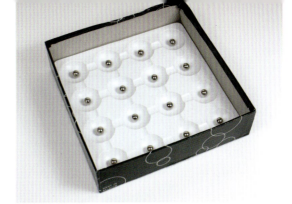
Quantum space

interested in showing how fantastic my imagination can be, but instead how we can look at our known world from a new and different angle, thus enlarging our perception. Physicists are doing the same, because they are looking for new ways to explain phenomena like gravity that we can observe every single day of our lives. Once while cycling home, I saw a shoe on the street, if I could see the shoe, I could also imagine a guy somewhere with only one shoe. While watching the billions of stars, physicists do not only see the light but also see the anti-matter of our universe, the hidden dark matter and so much more.

So I started to record the phenomena around me, it could be serious science but maybe it would only be absurd science. As I could not work on the scale of real particles I worked on objects that are made out of particles and started to use physics principles rather than explain physics. I attempted to grasp the reality of our world but I felt that I could also lose parts of it. I want to make things visible that we do not usually notice. This meant to play outside the margins of the known and the predictable in order to reach the unexpected! In physics an experiment has to be repeatable, but art can have an unpredictable result. This result may not be as exciting as those from the physics processes, but instead it was the experi-

ence itself, which counted. I found it very difficult to deal with theory, because the theories in physics are so beautiful but we are here on the ground and cannot see a quantum effect. Physicists are using mathematics as a way to model the world so I tried to use objects as a way to model the world, everyday objects. I started to make some basic experiments about particles (balls) and trajectories. It should look simple rather then tricky but play with the principal of moving the particles and creating the space around them. Soon many different models of space and particles were used to create other interactions. In the principal of relativity, space and time are tied together but people cannot for example actually see the speed of light. I rotated balls on moving surfaces, I built a machine to create random movements and I made drawings from collisions. To witness how atoms behave, I ground objects into powder – but I could still not see the atoms. I discovered the Coriolis force, but it was discovered almost two hundred years ago. So the physics I was making in my own lab: (the studio they so kindly gave me for my own experiments) was XIX century physics and XXI century physics used other devices. Through all the learning and experimentation, I had to accept that I could not make real physics, but I could create images and metaphors about it.

Minus mayo

Experiments on unpredictable movements with magnetism

Some examples include: a vacuum is never completely empty, it must therefore be possible to make a sculpture of the dust found in an empty room; a seed of a geranium plant becomes a flower, but what if it is ground and compressed into the shape of the seed again! If you divide A into X parts and multiply again by X you obtain A, but in reality, breaking an object into X parts and gluing those parts together again doesn't make the same object. This is the principle of entropy. What is a minus object in reality? If an empty tube of mayonnaise were equal to zero, what would be a minus quantity of that tube? Perhaps one can find out if the tube itself came out through its own hole? A star begins with dust particles joining together so if I built an installation with a circle of fans that holds dust of the room together would a young star be produced?

Conclusion

As an artist I think about our world and try to find possibilities to look at it, to see common things from a different point of view. Actually the physicists also do this but they tell us about what our world is made of. The aim of physics is the creation of knowledge through observation, reflection and experimentation. The aim of art is not as clear, but for me it is also a way of thinking about the world through observa-

tion, reflection and experimentation. As Robert Filiou said: 'Art is what makes life more interesting than art.' After this experience I would say: 'Art is what makes physics more interesting than art.'

I became interested in physics but can an artwork be simultaneously on the cutting edge of both the art world and the science world? I am not sure. Usually the science behind art is more interesting than the art. Perhaps the syndrome of the model is better than the artwork. Physics is about rational knowledge but art deals much more with how our perception depends upon our knowledge. In order to change our perception, we must shift our knowledge and by learning about contemporary physics I became aware of many things. I can see the world differently now.

The world has not changed, but I did. I cannot imagine a single artwork that I could make that would be able to summarize my swim in the pool of physics, so I decided to share my experience and the process of gaining my new knowledge through the medium of animation film. The Physics Department of Geneva University has also given me a commission to build a *Quantum Art Cloud* at the entrance of the University. This will be a module called the quantum *QUARC* and will showcase my researched experiments.

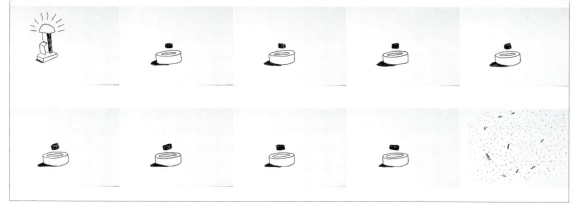

Drawings for the animations

Thanks

I would like to thank all the people who made this experience possible and shared it: Jill Scott, Irène Hediger, Martin Pohl, Olivier Gaumer, Ruth Durrer, Jean-Pierre Eckmann, Peter Wittwer, Didier Jaccard, Marie-Anne Gervais, Olivier Landry, Niels Madsen, Nathalie Chaduiron and Roland Pelet.

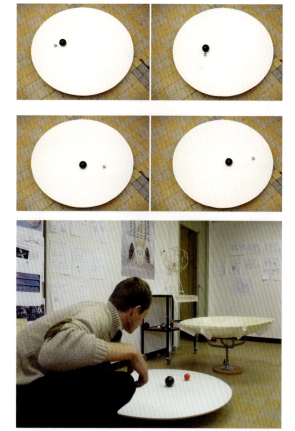

Experiments in the studio

Experimental set-up with parasol, July 2007

Roman Keller

ARTIST IN THE PAUL SCHERRER INSTITUTE (PSI), VILLIGEN

Disciplines: Installation, Photography, Video

Project title: *Energy Plan for the Western Man.* Art lives on hope! The availability of energy has taken a main role in everyone's day to day way of living. The search for alternative energy-technology has strong symbolic and emotional implications. The goal of this research was to analyse and attempt to visualise these implications.

ENERGY PLAN FOR THE WESTERN MAN Roman Keller

Admission by magnetic ID card only

My time as artist-in-lab at the Paul Scherrer Institute (PSI) began with a meeting at the Institute's Villigen HQ, where I was introduced to Beat Gerber, Head of Public Relations, and Fritz Gassmann, a physicist in the Energy Research Department. It was decided that I should present my ideas in a talk to that department two days later. My presentation – in which I focused on the ten day construction phase preparatory to an exhibition in the Kunsthalle Fribourg – was well received. I started out on my image research and gradually began to make contact with individual scientists. I did not have to wait long for the first disappointment: the PSI library had very little pictorial material connected with alternative energy sources – intensive research in this area only began at PSI towards the end of the 80s. But I did find (in the glossary of a six volume Encyclopaedia of Energy) a summary of the energy history of our planet, and this became the foundation for a new book project. My focus moved increasingly away from the original project submission, which was to produce the 'visual consequences' of alternative energy research, and I became engrossed in the idea of pursuing research of my own.

In the following weeks I worked on various concepts. A discussion with Fritz Gassmann on the different forms of motive propulsion led us jointly to the idea of solar-driven steam power. I had already conducted some work on steam-powered rockets, and together we conceived the project of developing the world's first solar-driven rocket. It was a winner – at last a real invention at the interface of engineering and art.

The basic idea was entirely simple: a flask of water with a conical jet at the base is placed at the focal point of a parabolic mirror. For safety reasons the opening of the jet must be remote controlled. The flask would be fitted with side fins for aerodynamic stability, a nose, a parachute, and an electronic signaling device for retrieving it at the end of its flight. With a measure of relief – and, as it turned out, at least as great a measure of naivety – I threw myself into the project. It would be a real PSI job, complete with budget, physicists and engineers, and I sent off my e-mails to Gerber and Gassmann, confident of their support. Then came the cold shower: six weeks later they called me to a meeting in which they informed me that PSI could not identify itself with the concept of a solar rocket. In the first place rockets were not a research focus of the Institute, and secondly they were associated with military use. They also doubted the feasibility of the project. I fell back to earth with a thud: this was the low point in our cooperation. After considerable soul-searching I decided to go ahead with my solar rocket without top-down support. Driven by a mixture of stubbornness and responsibility towards the artistic avant-garde I would reduce the project to its simplest possible shape. Perhaps admittance would now be without a magnetic ID card!

Named after the Swiss physicist Paul Scherrer, PSI was established in 1988 through the amalgamation of the Swiss Federal Institute of Reactor Research (founded 1960) with the Swiss Institute of Nuclear Physics (founded 1968). PSI is big; 1800 people work there, and when my placement started I knew no one. My office was situated in a decommissioned reactor building, where I slowly began to make contact with

Burning mirror by Peter Hoesen, 18th century

Steam rocket used for crash tests at Mercedes-Benz, 1962

other people. Without any action on my part, my next door neighbour, Robert Maag – a decommissioning engineer working on the first Swiss reactor – asked about my project. Slightly embarrassed, I told him of my idea of a rocket propelled by solar energy and the longer we talked about it, the more enthusiastic he became. It was the spark that ignited a real working cooperation. Robert supported me at every point, not only with his engineering know-how but in his awareness of how PSI functioned: one door after another began to open for me. When the rocket flies, it will be because he supported it so wholeheartedly – and also a little because I refused to take no for an answer.

Robert introduced me to the head of the apprentices' workshop, Marcel Dänzer, who took over the construction side of the project – a role whose importance grew all the time. His skill and involvement have brought the rocket where it is today. Once I started moving around the campus with bits of the rocket under my arm, people stopped to ask about it, and I had some interesting conversations not only with research scientists and engineers, but also with PSI managers and administrators.

Starting when the show's over

I finished my nine month placement with an artist's

logbook, a book about the history of solar powered vehicles and a presentation/exhibition of my solar rocket model. I intended to continue work on the project with my 'team' until spring 2008, in the hope that one sunny day the world – in the form of an interested group of PSI staff, along with invited friends – might see the launch of the first ever solar powered steam rocket. Behind my decision to use my artists-in-labs residency to develop a technological break-through lay the hope of using the methods and skills of scientists to revitalize a youthful dream: something that would awaken in them the original pioneering spirit of science and the emotional impact of discovery. It was in this historical and symbolic context that I started to explore. What remains with me from my time at the Paul Scherrer Institute is the insights this context gave into the working processes of Switzerland's biggest research institute. I learnt how to pursue my goals in a large organization whose ways of operating are very different from those of the world of art.

The most interesting aspect of working with scientists comes when comparisons can be drawn between the processes, methods and aims of art and science. In my case the interface with the technicians and workshop personnel at PSI was more intensive – a

Meeting with Roman Keller, Beat Gerber and Jill Scott at the Paul Scherrer Institute

Assortment of possible pressure tanks for the sun rocket

daily exchange of views far beyond the boundaries of my installation itself. Work on the project itself was difficult as I had to accept so many disappointments, but on the other hand the collaboration with the skilled PSI staff opened new horizons for me. Up to then I had been bound as an artist by my own personal limits; in this project I could profit from the immense knowledge and infrastructure of a high-tech institute. The extensive factual, historical and image research I undertook during the development phase gave me a great deal from the scientific context. My time at PSI was not a rest cure. In the course of the nine months I took part in six exhibitions, and my final balance was positive. I was able to start a project that will continue to occupy me for some time ahead. If it is to succeed, an artistic project undertaken within the framework of a scientific institution must set itself two goals: to make its mark in the life of the institution, and at the same time to be perceived as a serious contribution to the world of art.

Back to the future

The idea of stimulating an exchange between scientific researchers and artists is certainly in itself worthy of support. However, in practice the different viewpoints, expectations and incompetencies of the three parties to the project (the University of the Arts, the research institute and the artist) harbour a considerable potential for confusion. The artists-in-labs program should ensure that no false expectations arise on the part of the research institutes. Art cannot be instrumentalized for PR or other communications purposes. Artists must remain free in their decisions. Only towards the end of my placement did I begin to understand the extent to which I was living on another planet.

Special thanks to

Marcel Dänzler, Fritz Gassmann, Beat Gerber and Robert Maag.

Disused solar concentrator at the Paul Scherrer Institute

First experimental set-up with umbrella, May 2007

Installation view in Roman Keller's working and exhibition container at the Paul Scherrer Institute, November 2007

Experimental set-up for pressure test, September 2007

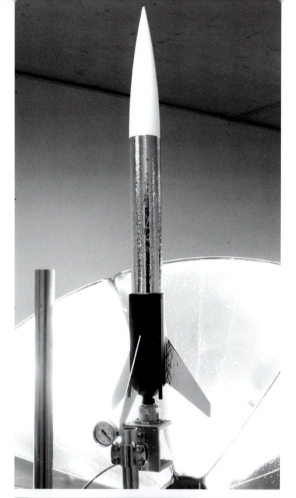

Final solar rocket by Roman Keller at the Paul Scherrer Institute, November 2007

Final kinetic speakers from *untitled_sound_project,* 2008

Pe Lang

ARTIST IN THE CSEM SWISS CENTER FOR ELECTRONICS AND MICROTECHNOLOGY,
ALPNACH

Disciplines: Sound Art, Electrician

Project title: *untitled_sound_project.* Analysing a room's attributes using acoustic analysis. Compiling new micro- and macrostructures in compositions based on the idea that the 'syntax' of music and/or composition is manipulated by an external system. Learning from new technologies and reinterpreting scientific phenomenon in an artistic manner.

UNTITLED_SOUND_PROJECT Pe Lang

March–May

I started work at CSEM two weeks before the official date, because the Divisional Head, Dr. Ulrich Classen, wanted to greet me personally on my arrival. He was in the process of leaving the research centre. He was not the only person changing positions at the time: Max-Erick Busse-Grawitz, Group Leader Sensor and Systems (Section 631), where I was to spend the next nine months, was also on the point of changing jobs. So, at the time of my arrival at CSEM as artist-in-lab, nobody apart from Marcel Gasser knew what the artists-in-labs program was about or why I was there at all. I rather enjoyed that.

Marcel Gasser became my main contact. He took great trouble to answer my questions and made sure that I had all I needed in the way of equipment by the end of my first day. My workplace was more like an office than a studio: no paint, stone, sculpture or other typical trademarks of the artist, but simply an adjustable table, a telephone and a Macintosh with Wacom and stylus. Maybe this is partly why Dirk Fengels, current Section 631 group leader, sometimes saw me as an engineer more than an artist. The artists-in-labs program is not a classical studio placement abroad but a protracted period of cooperation with scientists and engineers. The first three months were spent getting to know the incredibly wide range of CSEM projects and departments. CSEM is not a pure research laboratory: it works at the interface of research and industry. The website only gives a faint idea of what for me was an entirely new field of experience. I had to prepare myself to be open and receptive to new ideas and inspiration. CSEM Division A in Alpnach is home to some 35 scientists and engineers. Over the whole 9 months of my placement I was able to accompany current projects and was kept up to date on progress and problems. It was an exciting time, as I am always keen to explore new methods and technologies. I made a lot of contacts on my regular 'rounds' in the first few weeks, and we had many discussions, mostly of a technical nature. One advantage of a group of 35 colleagues is that there is always something to celebrate – birthdays, weddings etc. – so pauses for 'apéros' (social drinks) and other snacks were the rule rather than the exception. Weekly group meetings opened up new and fascinating perspectives for me. Solutions to current problems were discussed, schedules fixed, and individual progress reports given.

I have been an art autodidact ever since I qualified as an electrician; I never even considered taking formal training as an artist. Not that I sat down ten years ago and said 'Now I'm going to make art' – I just did what fascinated me and discovered whatever that brought. In that respect, a CSEM project plan was something new to me, but it was also meaningful. Clearly structured procedures of this sort really were unexplored territory for me, and they could have been a torment, but I set out to define my own project along the same strict lines in order to achieve a specific goal with a professional result. I did not want something 'makeshift', as the engineers would say. So there I was, creating my project plan and learning UML2 (Unified Modelling Language), a standardized language for modelling software and other systems.

A presentation of my earlier work brought the worlds of art and science closer together; in their technical

Pe Lang with a motor sample from Maxon Motor

Soldering station CSEM

realization my creations are, in fact, not altogether dissimilar to those of scientists. My interest in technology and my (albeit modest) knowledge of electronics helped to foster a profitable exchange of ideas. I put up a video film 'Der Lauf der Dinge' ('How things go') by Fischli/Weiss in the cafeteria; only two or three people had ever heard of them, and it was interesting to see how engineers approached such a work. My project plan stipulated that I had to decide on a specific project within the first three months. With Dirk Fengels I looked at numerous suggestions to see if they were feasible within the artists-in-labs framework. Dirk was always enthusiastic about my ideas, which was immensely encouraging. Everything seemed – and still seems – possible. Marvellous!

My method was to make a prototype early on in the process, so that I can see how it worked in the real world. I studied and experimented with a 3D camera, and with a delta robot and various applications, as well as with a number of innovative sensors and wireless networks. So-called 'demonstrators' were of particular interest, because they showed new technological functions simply and accessibly. I knew my project must have the immensely precise movement of such devices. I began to work on the software for

acoustic analysis, but soon realized that time did not allow for both aspects, so I adapted the focus of my original submission – an intelligent acoustic composition system – to an output system consisting of 8 turnable computer-controlled loudspeakers. Discussion with a number of colleagues in the lab brought some spontaneous new ideas about linking the movement of the speakers via 3D camera or WiseNETTM (low power wireless sensor networks). It worked!

June – August

Then the internet marathon started. It's amazing how much time goes by looking for, comparing and selecting the components one needs. It was important for me to keep within the predetermined budget, and another parameter was that I should be able to do the construction work myself, only calling on the scientists and engineers for help where necessary. I learnt a lot from Thomas Bruch about motors, drives, encoders and amplifiers. I had to examine the specification of motors and accessories to ensure that they fulfilled the various requirements for power, running speed, and precision. The gearing had to meet specific dynamic parameters, and the whole assembly should run as silently as possible. In the course of the discussions about components it also became clear to everyone that function was not the only impor-

Software check using *Max/MSP* software

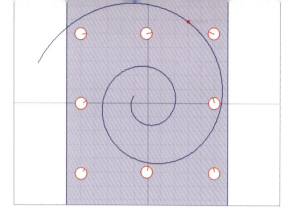

Graphical representation of how the sound travelled through speakers

tant aspect: the overall appearance of the installation was equally crucial. The initial prototype with an SAIA stepping motor and gearbox simply did not meet these requirements. The installation was within budget, but not up to CSEM standards.

Thomas Bruch then arranged for me to meet Mr. Schällibaum of Maxon Motor, which had built the motors for the 'Sojourner' Mars rover. I took the train to Sachseln and was surprised to see that the local station was called after the company. Mr. Schällibaum offered me a special price, and I began to see my way ahead again. I was allowed to test a flat motor whose noise specifications had not been recorded on the data plate, which brought me another step forward, and the pieces of the puzzle began gradually to fall into place. With Krzysztof Kransnopolski, the best fabricator I know, I started to redesign the project.

It was reduced in size, which was optically advantageous. The complex inner workings were now hidden, but a child could still see how the loudspeakers were moved. My workplace was bare: I had nothing to show, nothing that could be assessed, just a lot of data specifications for ICs, sensors, slip rings, cables, motors and so on. As there was nothing to see, there was at this stage nothing one could properly call a

reaction from the scientists or engineers. This must change, I thought. So I started to order the components I needed, in the awareness that I was going to be miles over the official budget ceiling. I had to turn to private sources and hope for external funding – unfortunately in vain.

Then something I had not accounted for happened: I had entirely forgotten that the summer holidays were due and, even worse, that the companies making the mechanical parts could only take new orders 6 weeks ahead. My whole timing was shattered.

September – November

The components began to arrive, and I spent all my time working on them, soldering 160 plugs and sockets – over 1000 connections in all – and preparing control and interface PCBs.

My workplace began to fill up, and people stopped by to ask questions. But the CNC-cut anodized aluminium elements I needed to put the product together still had not been delivered. My desk looked like an exploded drawing. After the calm, however, came the storm, and I brought a sleeping mat to work with me. There was often no good reason to stop work in order to catch the last train home (22:25), so I simply slept

3D models of the kinetic speaker

Turny mechanism for speakers

at CSEM. There was a lot to do. The aluminium parts arrived, and the only thing missing now was the 8 big gear wheels. Once the first loudspeakers had been assembled, the interest of scientists and engineers in the burgeoning work of art began to grow. The point and pointlessness of such an installation was the line along which art entered science, and I was often urged to patent the entire system. However, the professional assembly of the loudspeaker system brought the next problem. My original idea of linking the speakers and their kinetic control to the central computer with three separate cables seemed make-shift and improvised in comparison with the speaker assembly itself. Because of the expense of a single cable solution for each speaker unit – this would require a special type of cable. The upshot was that I found myself two days later en route for CSEM in my mobile workshop (Swiss Rail) carrying 100 m of special cable that weighed 25 kg.

Installing this brought my already limping schedule to a complete standstill. The plan was to have all 8 loudspeakers working by November 28, which would leave me some 50 hours in reserve to complete a brief composition and choreography for the demon-stration. Parallel to this I now developed the control algorithm for the motors to turn the speakers a pre-

cise 0.04°; it was a matter of luck that the software, except for a few bugs, worked perfectly. Initially the control software would function merely as a con-necting link between logic pro and the speakers: this was the simplest solution, and would leave me time to finalize my preparations in the remaining hours before kick-off. I soldered the cable into place, so that four of the eight loudspeakers would be ready for the demonstration.

Conclusion

I sometimes ask myself if my presence helps scien-tists to see their work from an artistic perspective (laughs). Maybe changes really were made in the design of robots or presentations because of me. Anyway, there was certainly an exchange of ideas. My nine months in a science company were first and foremost a meeting with people. I was often intro-duced as a closet engineer but at CSEM I met inter-esting and stimulating people who brought in new ideas and allowed me to transcend my own bounda-ries. I hope this was the same for them, too. Towards the end of the project I heard comments like 'Already over? So what happens next?' or 'Will someone else come?' Collaborating with CSEM was such a natural and normal experience that the idea of its ending met with a certain amazement all round. It might be

Inductive sensors

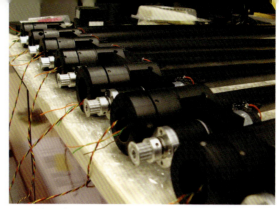

Assembling the 8 speakers

possible in the future to find a form of networking that would work for both parties, with their different structures, and enable further joint projects. At the end of those nine months, leaving was not easy.

Special Thanks to

The CSEM Team in Alpnach, the Swiss Federal Cultural Office, Sukandar Kartadinata, Heinz Schällibaum and Maxon Motor, Irène Hediger, Jill Scott and Rachel Bühlmann.

2 of the final models

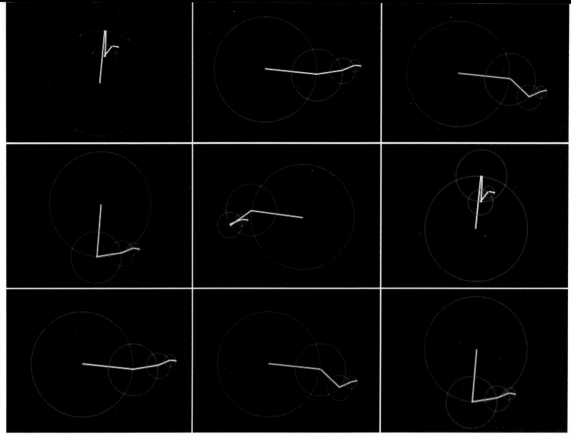

Images from *Chronophasis,* an installation developed using the *Diglossia* platform

Chandrasekhar Ramakrishnan

ARTIST IN THE NATIVE SYSTEMS GROUP | COMPUTER SYSTEMS INSTITUTE ETH ZURICH

Disciplines: New Media, Composer

Project title: *Di*glos*si*a* is a sociolinguistic term used to describe a situation when two languages, (often a spoken vernacular and a second written or spoken higher language) are commonly used by members of a single population. My proposal was to design an alternative bilingual program language and platform for developing multimedia software, which can combine the strengths of native environments with 'off the shelf' scripting languages (*Base, Max/MSP, Supercollider, Director/Flash,* and *Processing*). Using processes of integration and combination, the aim was to construct a platform for workshops for artists and scientists.

DIGLOSSIA Chandrasekhar Ramakrishnan

As a sound designer and programmer, I came into this residency with a fairly concrete picture of what I wanted to accomplish and what I needed to learn from the researchers. During the second week of the residency, I made a presentation detailing my work and its sound-art context, my influences, and what I intended to create in the nine months of the program. This presentation not only helped me communicate my perspective and ideas, but also served as my introduction to the Native Systems Institute and the members of other research groups within the same building. After this presentation I ended up working very closely with three members of the Native Systems Group: Sven Stauber and Florian Negele, and Dr. Felix Friedrich. I also developed close contacts with Nicholas Matsakis in the Lab for Software Technology headed by Dr. Thomas Gross. Overall, the scientists seemed to be excited to have me and I enjoyed the intellectual exchange. They enjoyed my different perspectives and approach.

The project idea of *Diglossia*, was to create a new platform for a particular kind of multimedia performances, designed to support artists and motivated by three primary concepts: 'algorithms, systems, and improvisation'. *Diglossia* was an attempt to bring these three concepts under one umbrella platform and I had to explain to the scientists why I regarded the construction of an experimental platform that uses these primary concepts to be a new artwork. Some people have the misconception that 'algorithms' are a product of the computer and the computer age, but this is not the case at all. An algorithm is simply a series of instructions for doing something. Some artists use algorithms very self-consciously,

such as the members of Les Algoristes. This is, however, not the only way algorithms are put to use in the creation of art. For example, there are aspects of conceptual art that are algorithmic in nature, like Sol LeWitt's wall drawings. The wall drawings are a series of instructions to be carried out to create the drawings. They were never drawn by LeWitt himself. He only created the instructions, the algorithm, that, when carried out, produces the work.

My interest in 'systems' comes from a cybernetic approach: that is, a process that controls or regulates itself. The use of systems has a long tradition in electronic music and composers like David Tudor, Alvin Lucier, and Pauline Oliveros have spent many years designing instruments based on cybernetic systems. Many musical and theatrical traditions around the world have incorporated 'improvisation' as a central part of the performance culture. Jazz music performances, for example, are often not scripted, but there might be a frame, a melody and set of chords, that defines the boundaries. The actual performance incorporates improvisation in its realization: the composition is simply a sketch and requires a set of performers to complete it.

Diglossia

Diglossia was an attempt to give the performers a set of necessary tools in order to define some algorithms and systems, and to improvise with them. This could be done either within their prescribed bounds or by altering the definition of the algorithms and systems during a performance. In this capacity, the scientists felt that the platform inhabited a strange place between art and computer science. The sources of its

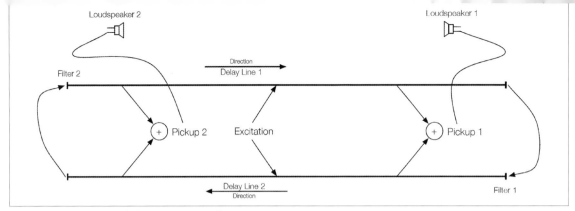

Schematic of a simulation of Alvin Lucier's *Long Thin Wire* instrument

for new content. As the crisis in the world financial system had reached a particularly critical point, I started to try and understand economic concepts that were used to describe this crisis. I was very intrigued by the similarities between the financial chaos and cybernetic systems. However in the end I opted for minimalist visual forms. As Sol LeWitt states: 'The form itself is of very limited importance; it becomes the grammar for the total work.' I felt that the basic form should be deliberately simple so that the form may more easily become an intrinsic part of the entire work. Using complex images only disrupts the unity of the whole, while repeating simple ones concentrate the intensity on the arrangement of the forms and their relational aspects. Therefore, I chose a white background and circles of varying size. Artists could improvisationally and algorithmically control thickness and aspects of the circles, their position, size, or movement, in *Diglossia*. These concepts were new for the scientists. I began simulating Alvin Lucier's *Music on a Long Thin Wire*, on a long thin wire instrument and using the *Diglossia* toolset to control it. In the end the manipulation of black circles on a white background can shift the sounds generated by a simulation of a long, thin wire and *Diglossia* was used to control, manipulate, and arrange both the sound and the visuals.

Conclusions

Immersion inside the Native Systems Group of computer scientists has allowed me to devise a new toolset that I will continue to utilize and improve in the future. *Diglossia* will certainly form the basis of future works and the experience has had a profound and surprising influence on my sound work. Not only have I gained more technical knowledge, but I have also gained new insight into the ways ideas can be communicated. I hope the computer scientists have also profited from me and my different ways of thinking and that they have enjoyed seeing the tools of computer science applied to very different situations than those of their trade. For me, the more important thing is that those users of computer technology can include artists and musicians in order to help define a vision of where the technology can lead us to in the future. Artists might not share the same visions as computer scientists, but our approaches are just as valid and compelling. So far an initial version of *Diglossia* has been completed and was demonstrated in a performance on the 19th December 2008, at the Zurich University of the Arts (ZHdK). At the end of my residency, Professor J. Gutknecht and the ETH Zurich hired me to continue working on their environment for another year, an opportunity for which I am very grateful.

Schematic drawing of Alvin Lucier's *Long Thin Wire* instrument for an installation

Special Thanks to

Prof. Gutknecht and the Native Systems Group for providing me such a hospitable environment to realize this work; Jill Scott and Irène Hediger for their support; and James McCartney for the stimulating discussions and his invaluable suggestions.

Chandrasekhar Ramakrishnan explains *Diglossia* (video still)

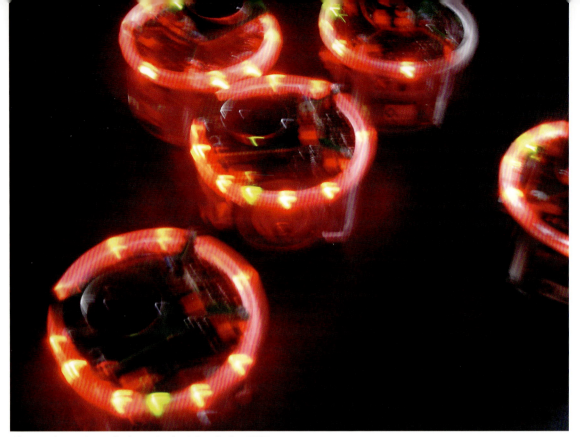

When I woke up, the sun had moved, robotic installation, 2009

Alina Mnatsakanian

ARTIST IN THE ISTITUTO DALLE MOLLE DI STUDI SULL'INTELLIGENZA
ARTIFICIALE (IDSIA), MANNO-LUGANO

Disciplines: Conceptual Art, Painting

Project title: *When I woke up, the sun had moved,* an installation using 10–12 small robots that
followed a choreography and 'danced' with each other and/or with human 'partners'. This project is a formal
continuation of *Post Minimal Sculpture.*

WHEN I WOKE UP, THE SUN HAD MOVED Alina Mnatsakanian

Month one – March:
Fascination and Learning

The first month was a period of discovery and fascination but there was no formal process. The head of the robotics lab, Alexander Förster introduced me to the robotics lab; we discussed my project and he suggested that I use a set of e-puck robots. Coincidently, the e-pucks were designed at EPFL by Francesco Mondada with whom I had collaborated on an earlier installation in 2007 entitled *The Mountain comes to me*. This time I really wanted to learn to program these robots, so Alexander suggested that I start with C programming language and gave me a PDF tutorial. The rest of the month was divided between trying to learn the C language and talking to IDSIA researchers to find out about their own fields of research. It was fascinating to learn about the background concepts of Artificial Intelligence and I was often referred to relevant books and papers which I started to order from the library. Frederick Ducatelle, a post-doc researcher working in robotics, suggested an interesting book by Valentino Braitenberg called *Vehicles*. Reading it gave me some insight into the 'behavioural abilities' of robots. At the same time I participated in Luca Maria Gambardella's Masters class at the University of Lugano (USI), which gave me some idea about swarm robotics and optimization and I also participated in Alexander Förster's class and the robotics lab.

Month Two – April:
The Simulation Program

At the beginning of April I organized a meeting with the two directors of the institute, Luca M. Gambardella and Jürgen Schmidhuber, to get some more direction and feedback. Luca suggested that I use a simulation program to create my choreography before starting to work with the robots. Both Alexander Förster and Frederick Ducatelle suggested that I learn the *Webots* simulator, a software with a graphical interface, developed by Cyberbotics Ltd. I mostly learnt this program on my own with a manual because the researchers at IDSIA actually use another, more complicated simulator rather than *Webots*. I easily grasped the graphical part of the program and soon I was able to create simulated clips of a few steps that were later used to code in the C language.

At the same time I continued my research into AI concepts and robotics. Most work in robotics had a basic goal or a behavioural reason and in most cases the aim was to display the capabilities of the robots. Of course robots do not acquire intelligence, but they receive 'intelligent' programs and behave accordingly. They might look like autonomous agents, but the lines of code and the mathematical algorithms actually dictate their behaviour.

Month Three – May

Month of May was a month of stagnation. I really felt the need for support and that I was not progressing well. The technical hurdles did not allow me to think in a very creative manner and I became very frustrated. At the same time I started thinking about choreography. *Webots* were a feasible means to think about movement when I used the visual interface, but when I reached a point that I had to use C programming, I simply did not know enough. When I asked for more help to learn C, the scientists did not have time to help me. So I took some refuge in creative music and

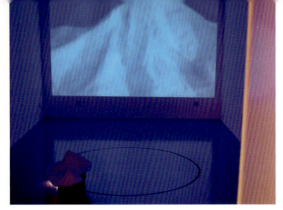

The mountain comes to me, installation with 5 videos, a 3D model and a robot, 2000

With Jérôme Guzzi at the IDSIA robotics lab (video still)

composed by placing bits of music samples together, first in my original idea of percussions from various cultures and then in a more sophisticated way using various sounds.

Month Four – June: Official Support

Greg Kaufmann, a Masters student in Intelligent Systems at the University of Lugano (USI) started to help me and we had weekly brainstorming sessions during the month of June. With his suggestions I started to learn C again, using a simpler book, but still working on my own. By using the *Webots* I created the steps to be used in the code, and I continued with my own digital musical compositions. I attended talks by the IDSIA researchers in the domain of robotics and AI, as well as the Master's students' project presentations. A group of IDSIA researchers initiated a program called *Appetizer*, which are weekly half hour presentations before lunch. Although most topics were difficult to follow participating in them kept me in the loop of IDSIA research. My general impression was that tasks that might look simple to a non-connoisseur are quite difficult to program and require months of work. Furthermore, robots do not always behave as the program predicts, and environmental changes can have unwanted or surprising results in many experiments.

Scientist-in-Studio

Being fascinated by the international cast of IDSIA, I initiated a side-project called *Would you be my scientist-in-studio?* I asked IDSIA colleagues, via email, if they would like to be my scientist-in-studio and answer my question: 'Which is YOUR mountain?' This project was a thematic continuation of an older work *The mountain comes to me,* but at the same time it was also my way of establishing a closer relationship with the scientists. Nine scientists have replied so far. I have recorded one person and will continue with the project, either by interviewing everyone, or by creating an online installation or a Skype performance.

Month Five – July

Greg Kaufmann was recruited at IDSIA for a summer internship and I followed the progress of his work. Unfortunately he did not have any time to help me with my project. I started working on clay models to cover the mobile robots. These consisted of either abstract organic shapes, hats or houses. This activity required special attention because each e-puck robot has eight infrared sensors and a camera, which constitute the main means of its communication and should not be covered. During this part of the research I thought of a different name for my installation, in relation to my artistic research on identity:

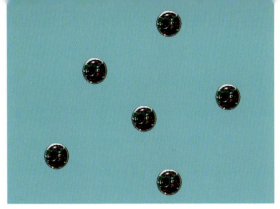

A simulated scene for the choreography (video still)

A sketch with robots balancing hats

'When I woke up, the sun had moved,' instead of the original name 'Dancing Robots'.

Month Six and Seven — August–September

Frederick Ducatelle, volunteered to help me with the programming. He had already assisted me by choosing the simulator and explaining some general concepts, so he created a base of modular codes that were based on my steps. After this I was easily able to manipulate the numbers, create my choreography and experiment with the robots movements. Everything suddenly became much more interesting. I started looking at the robots as real dancers with specific morphologies. A small wheeled robot that has a cylindrical shape and moves using its two wheels and motors looks better performing certain steps than others. Also the same music that sounded appropriate for the simulated robots was no longer suitable for the real ones. Also, unlike simulated robots, the real ones made a lot of noise, so I thought that this noise should be used as a sound element for the actual sound track.

Month Eight and Nine October–November

Another interesting discovery was to find out about new software that had been developed by Alessandro Giusti, an IDSIA researcher and one of his students. Although the software is for tracking movement for security and detection purposes, I immediately saw the potential to use it to track the movement of the robots. Alessandro agreed to work with me. I continued to fine-tune the choreography and concentrated more on the shapes and movements of the robots and also used their lights, in order to create an atmosphere. For a longer battery life in future exhibitions, I wanted the robots to be triggered by sound and after performing their movements go into sleep-mode. Jérôme Guzzi, an IDSIA researcher, liked the idea and tried it. I spent more time in the lab to test the concept. We also tried Alessandro's tracking software and everyone was happy with the results.

In December I was able to invite everyone to a presentation of my installation at the laboratory. My collaborators were happy with the results and showed interest in continuing to enhance the installation even after my residency was officially finished.

Thanks

Very special thanks to Frederick Ducatelle, Jérôme Guzzi and Alessandro Giusti, IDSIA researchers, as

Clay models of hats

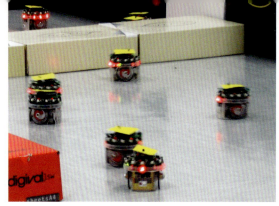

An experiment with e-puck robots

well as Luca Maria Gambardella, director, IDSIA, Alexander Förster, senior researcher and Greg Kaufmann, Masters student. Thanks to all IDSIA researchers and staff for their insights, friendship and sense of humour. And, of course, many thanks to Jill Scott, Irène Hediger and Karin Rizzi of artists-in-labs program Switzerland.

E-puck robots in motion

Close-up of robots

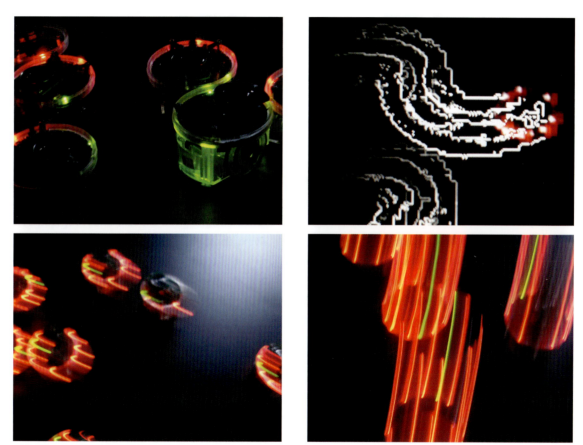

This series shows how the movement of the robots generated a choreography of light patterns

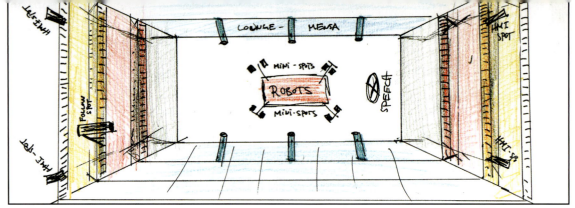

Sketch for the AI Lab 20th anniversary celebrations installation *Kinetic Spaces,* 2007

generated choreographies that were built in Artificial Life with Daniel, could control the humanoid dancing robot. This research became the foundation for a 'Choreographic Machine' project which would involve the humanoid dancing robot to be choreographed by the *ALife System*. Also at this time I attended an excursion with the lab to the Wearable Computing Lab at the ETHZ, and was impressed by development of costumes with sensors and actuators, which could register the positions of the articulations of a human body. Perhaps their motion tracking could be tied into our new *Life Forms* software? However, such a project would require another 9 months of artists-in-labs residency for its development.

September–November 07

Throughout the month of September the first designs of the robot leg were built using parts produced by the in-house 3D printer. Discussions about locomotion and articulation led to another prototype. Rolf Pfeifer commissioned a performance called *Kinetic Spaces* for their 20th anniversary celebrations. I had recorded sounds and images throughout my stay in the AI Lab's premises, so I manipulated and amplified the daily sounds that go unnoticed in such premises, and combined these with videos of robots I had made at the AI Lab, including a recording of the

choreography of my *Kondo* robot. This commission included the design of the installation, the lighting of the space with a professional lighting designer, the realization of a video and audio design and the setting of a time line of events to take place during the presentation/installation. A robot head acting as master of ceremonies introduced the various speakers that were to participate in the anniversary celebrations. A script was written by the lab members for the robot head to 'speak'. Through this collaboration with me, I think they attained a first-hand experience on how an artistic event and performance can be organized and staged as a dynamic spatial event.

Conclusion

My residency opened me up to the possibilities for new projects, provided contacts for future collaborations and enabled me to be immersed in a scientific field of AI. It provided me with the possibility to learn Artificial Life programming and to apply this to my work. Finally it helped to bridge a gap between scientists in the lab and myself through collaborative efforts. While 9 months proved to be an adequate time span to invest in such an endeavour, it is a very short time to conceive and build such an ambitious artwork as a humanoid dancing robot. But the collaboration with the individual engineers directly related

AI Lab 20th anniversary celebrations (2007)

Dancing robot foot (Image: Raja Dravid)

to this project will hopefully continue. The support for such an art/science project may not come from the AI lab, because projects like these do not sit with their research priorities. Who would own the final prototype? At this point they officially assumed this project would be one of their doctoral thesis topics, but surely an artist must be in charge of the way it moves.

Credits

Many thanks to Rolf Pfeifer, Daniel Bisig, Raja Dravid, Max Lungarella, Jill Scott, Irène Hediger and everybody at the AI Lab who took part in the *Kinetic Spaces* Installation.

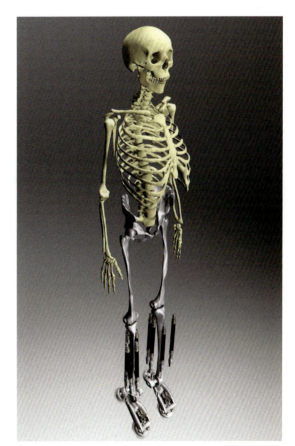

Humanoid dancing robot for the *Choreographic Machine* project (Image: Raja Dravid)

BIOGRAPHIES OF WRITERS AND ARTISTS

BIOGRAPHIES OF WRITERS AND ARTISTS

artists-in-labs Research Team

JILL SCOTT was born in 1952, in Melbourne, Australia and has been working and living in Switzerland since 2003. Currently she is Professor for Research at the Institute for Cultural Studies in the Arts, at the Zurich University of the Arts (ZhdK) in Zurich and co-director of the artists-in-labs Program (a collaboration with the Swiss Federal Office of Culture). She is also Vice Director of the Z-Node PhD program on art and science located in both, Zurich and at the University of Plymouth, UK. Her recent publications include: *artists-in-labs: Processes of Inquiry* (Springer, Vienna/New York 2006), and *Coded Characters* (Hatje Cantz 2002, Ed. Marille Hahne) and she has written many chapters in international books on new contexts for the arts. She has a PhD from the University of Wales (UK), an MA from USF, San Francisco and two undergraduate degrees (in education and in art and design). Since 1975, she has exhibited many video artworks, conceptual performances and interactive environments in USA, Japan, Australia and Europe. Her most recent research for her artworks is about neuroscience, embodiment in the environment and brain plasticity involving the construction of interactive media and electronic sculptures. <http://www.jillscott.org>

IRÈNE HEDIGER is co-director of the Swiss artists-in-labs program at the Institute for Cultural Studies in the Arts (ICS) at Zurich University of the Arts (ZHdK) in Zurich. She is also head of the China/Swiss Residency Exchange – an intercultural exchange project on art and science (a collaboration with the Swiss Arts Council Pro Helvetia). Other activities at the ZHdK include her role as Deputy Equal Opportunities Officer. After her studies in Business Administration, she got a degree in group dynamics and organizational development (DAGG) and a Masters of Advanced Studies in Cultural Management from the University of Basel. Irène has curated numerous exhibitions and cultural events in Switzerland. Internationally she curated the exhibition *Lucid Fields* at ISEA (International Symposium on Electronic Arts) in Singapore in 2008 and currently the travelling exhibition *Think Art – Act Science* in Barcelona, San Francisco and other destinations (2010–2011). She specializes in inter- and transdisciplinary creative processes and practices and the development of inclusive and participatory outreach concepts for a general public. Irène is part of a think-tank on new media arts and is regularly invited to expert meetings and to workshops. She has written essays on interdisciplinary practices.

MARILLE HAHNE is a filmmaker and Professor of Film, and co-director of the Swiss National Masters program in Film Directing at the Zurich University of the Arts (ZHdK). She is a researcher in the field of HDTV digital cinema <http://www.digitalcinema.ch> and in 2005, she edited a book entitled *Digitales Kino – Filmemachen in Highdefinition mit Fallstudie (Digital Cinema – Filmmaking in HD including case studies)* <http://www.schueren-verlag.de>. She specializes in the blend between art, science and technology and is a free-lance editor for text and the kinetic image. She is the co-founder of the artists-in-labs project and director of the DVD documentaries produced within this project. In 1978, she completed a degree in Precision Tool Engineering und Optics at the University of Applied Engineering in Munich,

Germany. In 1980, she was awarded a Masters in Fine Arts in Film at the School of the Art Institute, Chicago, USA. Since then, she has produced numerous experimental and documentary film works and these films (e.g. *Augenlust*) have been shown at international film festivals. Her workshops include documentary video workshops in Hyderabad in collaboration with the Max Mueller Bhavan Institute and courses at the University for Television and Film in Munich.

Invited Writers

ANDREA GLAUSER is a post-doctoral researcher and lecturer at the Department of Sociology, University of Berne. Her main areas of interest include the sociology of arts and culture, social theory, history of social sciences, and qualitative methods. She has published articles on art and spatial theory and is author of the book *Verordnete Entgrenzung. Kulturpolitik, Artist-in-Residence-Programme und die Praxis der Kunst* (transcript, Bielefeld, 2009) and co-editor of *Der Eigensinn des Materials: Erkundungen sozialer Wirklichkeit* (Stroemfeld, Frankfurt am Main/Basel, 2007). She has studied sociology, art history, philosophy, and economics and holds a doctoral degree from the University of Berne (2008). In 2006, she was a visiting scholar at Columbia University in New York.

LLOYD ANDERSON is Director of Science at the British Council. He went to Highgate School in London, read for a BSc in Botany at Imperial College, and then studied for a PhD in Plant Sciences at Lancaster University. Since then he has held post-doctoral fellowships (at Manchester, Edinburgh,

Cambridge and Bangor) and worked for the Natural Environment Research Council. He joined the British Council in 1998, and is responsible for setting corporate policy in science, in close consultation with key UK stakeholders and partners, and helping to translate that global policy into strategies at the regional and local level.

DOMINIK LANDWEHR has always been fascinated by the inner life of electronic gadgets and machines, as well as by the stories and myths that surround them. He holds a master from the University of Zurich in German literature and cultural science and a PhD in media science from the University of Basel. In his promotional thesis he examined the myth surrounding the cipher machine Enigma. He worked as a journalist, radio and TV producer; and as a delegate for the International Committee of the Red Cross he served in various crisis areas such as the Afghan border. Today he is head of the pop and new media department of Migros Culture Percentage in Zurich where he manages and oversees a number of projects such as the platforms <http://www.digitalbrainstorming. ch> or the youth contest <http://www.bugnplay.ch>. Landwehr regularly publishes articles on the history of computers and technology. He also teaches media history at the University of Basel.
<http://www.peshawar.ch>
<http://www.sternenjaeger.ch>
<http://www.mythos-enigma.ch>
<http://www.kulturprozent.ch>

ROY ASCOTT is an artist whose work is invested in cybernetics, technoetics, telematics, and syncretism. He is the founding president of the Planetary

Collegium, based at Plymouth University, and has held senior academic positions in the US, Canada, Austria and the UK. His international exhibitions range from the Venice Biennale to Ars Electronica, with retrospectives in England and South Korea. His theoretical work is widely published and extensively translated. His has advised media art institutions in Europe, Australia, South America, the USA, Japan, and Korea. He edits *Technoetic Arts: a journal of speculative research* (Intellect) and is Honorary Editor of *Leonardo* (MIT Press).

NINA ZSCHOCKE is a post-doc researcher in the Department of Art History at the University of Zurich, where she is responsible for the scientific coordination of the PhD program ProDoc Art&Science (SNF), a collaboration of the Universities of Zurich, Berne, Lausanne, Fribourg and Geneva. She was born in Cologne, Germany in 1974 and has studied art history, ethnology, and classical archaeology in Cologne and London. She worked as a research assistant at MARS: Exploratory Media Lab (IMK, Fraunhofer Gesellschaft, Germany) and in 2000–2002 received a research grant by the Graduiertenförderung NRW. Nina Zschocke has earned her PhD from the University of Cologne in 2004 (Professor Antje von Graevenitz). Her book *Der irritierte Blick: Kunstrezeption und Aufmerksamkeit* appeared in 2006 (Munich: Wilhelm Fink). Since 2005 Nina Zschocke is located as a Post-Doc at the University of Zurich where she has taught several courses. 2006–2008 she received a research grant by the Deutsche Forschungsgemeinschaft (DFG) and was a visiting scholar at UCL/London and at Columbia University/New York City. Nina Zschocke has published essays in various journals.

Currently she is, together with Anne von der Heiden, editing a collection of essays by a number of authors related to ProDoc Art&Science and she is currently writing a book examining art history's relationship to neuroscience.

AURELIA MÜLLER studied art history, media and communication sciences and management at the University of Berne, Switzerland. She graduated in 2006 with a Major in Modern Art History under the supervision of Prof. Dr. J. Schneemann at the University of Berne. She was granted a scholarship by the Dr. Joséphine de Kármán Foundation of the Faculty of Humanities of the University of Berne for her 'Lizenziat' paper. During her studies, she also completed an advanced training in photography. She works in the field of national public support of art and design at the Federal Office of Culture (FOC) since 2006, supervises the public support of digital media art (project *Sitemapping*) and oversees on behalf of the Federal Office of Culture the federal support to the Swiss artists-in-labs project, among others. In the field of design and photography, Aurelia Müller acted as a curator for exhibitions at the *Museum für Gestaltung Zürich,* at the *Museum Bellerive,* Zurich, at the *Musée de design et d'arts appliqués contemporains* (mudac) in Lausanne and at the *Kornhausforum* in Berne. She has published texts about design, visual arts, digital media art, public support of art in Switzerland and the Swiss Federal Design Awards.

Artists in order of appearance

HINA STRÜVER has a Masters in Art from the Braunschweig University of Art in Germany, with a

focus on sculpture, installation and performance. After graduating she has had many international art projects and grant awards.
<http://www.hina.de>

MÄTTI WÜTHRICH studied environmental science at the Swiss Federal Institute of Technology in Zurich (ETHZ). Since then he has been an active political campaigner and intervention artist.
<http://regrowingeden.ch>

SYLVIA HOSTETTLER was born in 1965. She resides and works in Bern, Switzerland. She participated in various exhibitions at home and abroad and received numerous awards. During a journey in 2005, Sylvia Hostettler developed the first chapter *Luxflabilis* of the project series *Landschaften (Landscapes)* while on a discovery tour of the forests of the Lower Engadine Valley in Switzerland. With the artists-in-labs residency she was integrated in a research team at the Center for Integrative Genomics (CIG) of the Lausanne University in Switzerland for 9 months in 2008. There, she developed the fourth chapter of Landscapes entitled *Light reaction – Dimensions of apparent invisibility*.
<http://www.sylviahostettler.ch>

CLAUDIA TOLUSSO finished her Masters in Stage and Costume Design in 1999 at the Staatliche Academy of arts in Stuttgart under Prof. Jürgen Rose. From 2000 on she worked for 5 years as a professional freelance designer in Europe for theatre, film, museums and exhibitions, before she was invited to develop the performance design degree of the New Zealand Massey University and Toi Whakaari Dance and Drama School. In 2007 she created and co-ordinated with her colleagues the New Zealand student exhibition for the Prague Quadrennial International Theatre Design OISTAT. In 2009 she received the artists-in-labs stipend at the WSL research institute in Bellinzona resulting in a mandate from the Cultural Department of the city of Bellinzona to develop her three interactive installations in April 2011. In 2008 She was a foundation member of the art collective <http://www.1visible.net>. She continues developing set designs for different projects such as *JeruVille* in Düsseldorf and Hamburg and is a design tutor at the Art School Nendeln in Lichtenstein. Her future plans are to develop and bring arts into a bigger economical and social exchange.
<http://www.tolusso.com>

PING QIU was born 1961 in Wuhan, China and she studied there from 1981 to 1987 at the Kunstakademie in Hangzhou. From 1988 to 1994 she attended the Hochschule der Künste (HdK) in Berlin and since then she has been the recipient of many international grants, including the Erasmus Austausch-Programm (HdK Berlin-Royal College London), the Pollock-Krasner Foundation grant (New York and PRO HELVETIA, Schweizer Kulturstiftung Switzerland.) She has also received support for her work in Germany including the Künstlerhäuser Worpswede, the Institut für Auslandsbeziehungen Hauptstadtkulturfonds and the Akademie der Künste in Berlin. Since 1994 she has exhibited widely on a national and international level in China, the UK, Germany, Canada, Hongkong, Italy, Switzerland, Cuba and the USA. Some of her most recent exhibitions include an installation at the *Gallerienrundum in der August-*

strasse, Berlin, *the Chinese Arts Centre* in Manchester, UK and *Tier+ Mensch* exhibition at the Germanisches Nationalmuseum in Nürnberg as well as *Birthrites* at the *Science Centre Glasgow.* In 2009 her work was featured in *Transparenz* at the *Künstlerhäuser Worpswede,* Germany, *the Blackburn Museum & Art Gallery,* UK, the Museum of Contemporary Art, Skopje and at the *Celler Werkstatttage* in Germany. <http://www.ping-qiu.com>

LUCA FORCUCCI is a composer, performer and sound artist. He focuses on the sonic properties of architectural spaces, immaterial architectures and natural resorts. In 2010, he works at Cité Internationale des Arts and conducted research at GRM in Paris. In 2009, he worked at the *Brain Mind Institute* of EPFL, Lausanne, Switzerland. At the same time he started a PhD in the UK exploring sound and perception. In 2008, he travelled with *Francisco Lopez* in the Brazilian Amazon, for field recording. During the FIH 2008 (International Theatre Festival of Belo Horizonte, Brazil), he worked with independent media choreographer *Johannes Birringer.* Luca graduated in 2007 with a Master in Sonic Arts from the Queens University of Belfast. In 2001, *Al Comet* from *The Young Gods* produced his first album. Released at the Montreux Jazz festival by Universal records. His work was presented, among other places, at: FILE – Electronic Language Festival – Sao Paulo / Brazil; FIH, International Theatre Festival – Belo Horizonte / Brazil; Montreux Jazz Festival; Institut Suisse – Rome; Institut Français – Ljubljana; Stubnitz – Rostock; Fringe Festival – Edinburgh; Knitting Factory – Los Angeles. <http://lucalyptus.com> <http://www.myspace.com/lucalyptus>

MONIKA CODOUREY was born in 1966, in Warsaw, Poland. She lived and worked as an architect in USA, Canada and Germany before settling down in Switzerland in 1998. Currently she is a PhD Research Fellow at Z-Node of the Planetary Collegium and Institute of Cultural Studies located at Zurich University of the Arts (ZHdK) and University of Plymouth (GB). In her doctoral research she explores airport territory as interface and develops a human-oriented approach to design of hybrid spaces supported by mobile and social technologies from trans-disciplinary perspective (Architecture, Human-Computer-Interaction and Psychology). Monika holds her degree in Architecture form the University of British Columbia (CAN). She pursued her postgraduate studies in the field of Information Architecture at Department of Architecture, ETH Zurich (CH) and participated in VI Bauhaus Kolleg *Transnational Spaces* in Dessau (D). Monika has lectured in the field of Urban Media and Information Spaces at the Faculty of New Media, Zurich University of the Arts (ZHdK). She also was a guest lecturer at University of Applied Sciences Northwestern Switzerland, School of Arts and Design in Basel (CH). <http://monika.codourey.info> <http://www.constanttraveller.info> <http://www.mobile-identities.info>

CHRISTIAN GONZENBACH was born in 1975, he works and lives in Geneva. His practice is concerned with the transformation of ordinary things. Creating video, installations and sculptures, he works by making slight changes to daily raw materials in order to track the uncanny nature of day-to-day surroundings. He is exploring the boundary where our world may turn into the absurd or poetic by loosing

its rational meaning. Recent exhibitions includes Contemporary art centre in Versailles, Abbatiale of Bellelay, Musee Maillol in Paris, Fine art museum Le Locle. He is reprensented by Gallery Magda Danysz in Paris, Gallery SO in London and Gallery SAKS in Geneva. He won several awards during the last years. He is also teaching at Geneva University of Art and Design and is regularly invited as a lecturer at the Royal College of Art in London.
<http://www.gonzenbach.net>

ROMAN KELLER was born in Liestal, Switzerland and has a Master's degree in Environmental Sciences from the Swiss Federal Institute of Technology, ETH Zurich. He then trained as a photographer in Zurich, New York and Karlsruhe. Since 1997 he works as a visual artist, since 2003 most of his work occurs in collaboration with Christina Hemauer. He lives and works in Zurich. Christina Hemauer and Roman Keller have investigated the concept of energy for several years. One of their main areas of interest is the history of oil and its competing alternatives, notably solar energy. Thus the artist duo heralded the era of 'postpetrolism' for the arts with a manifesto and a performance in the year 2006 <http://www.post-petrolism.info>. The following year, their video installation *A Curiosity, a Museum Piece and an Example of a Road Not Taken* recalled Jimmy Carter's early and ultimately futile efforts to promote alternative forms of energy generation as symbolised by the installation of solar collectors on the roof of the White House.
<http://www.romankeller.info>

PE LANG, born in 1974, Switzerland, is a sound and visual artist based in Zurich and Berlin. His work includes sound installation, performance and composition. Lang has performed and exhibited his work in a number of festivals including Transmediale Berlin (GER), Elektra Montreal (CAN), Sonic Acts XII Amsterdam (NL), World New Music Days (SWE) and Galleries such as bitforms Gallery nyc (USA), isea 2008 Singapore (SIN), Växjö Art Hall (SWE), Netherlands Media Art Institute Amsterdam (NL), Cybersonica London (UK), Hatton Gallery Newcastle (UK). His work focuses on minimal and elegant kinetic systems, combined with different materials, which are used as sound sources. Selected awards and residencies include the *Sitemapping*/Mediaprojects Award, Bundesamt fur Kultur (2005 and 2008), Swiss Art Award (2009), artists-in-labs (2007) Center for Electronics and Microtechnology.
<http://www.pelang.ch>

CHANDRASEKHAR RAMAKRISHNAN is a media artist, composer, and multimedia engineer. His work, which exists at the intersection between conceptual art and free jazz, explores the interaction between algorithms, cybernetic systems, and improvisation. He earned a B.A. in Mathematics from the University of California, Berkeley, and an M.A. in Media Arts and Technology, with emphasis in electronic music and sound design from the University of California, Santa Barbara, where he studied with Curtis Roads and Stephen T. Pope. In 2003–2004, he was a fellow of the Akademie Schloss Solitude in Stuttgart. There, he collaborated with Max Neuhaus and several other fellows to realize Neuhaus' Auracle, a distributed sound installation on the Internet. From 2004–2008, he was a researcher at the Zentrum fur Kunst und Medientechnologie (ZKM), Karlsruhe. At ZKM he helped develop the Klangdom, a concert sound spatialization system, and developed the Zirkonium software, which was awarded second prize in the Lomus 2008 open-source music software competition. He lives in Zurich, Switzerland with his wife and son.

ALINA MNATSAKANIAN, a conceptual artist and painter of Armenian origin, was born in 1958 in Tehran, Iran. After studying architecture at the University of Tehran, Mnatsakanian moved to Paris, France where she received her B.A. in Visual Arts at the University of Paris. In 1984 she moved to Los Angeles, where she received her M.A. at the California State University, Los Angeles in 2000. She began the program as a painter, but soon ventured into installation and video. During this time she also developed an interest towards the exploration of identity through art. How does one feel about the multicultural influences? Where is home? Since 2005 Alina Mnatsakanian lives in Neuchâtel, Switzerland. She exhibits her installations and paintings internationally. Her interest in technology took a turn in 2007, when she received a production grant from the Swiss federal Institute of Culture for an installation with 5 videos and a robot, which was made through collaboration with the robotics department of the *École Polytechnique Fédérale de Lausanne (EPFL)*. In 2009 she followed her robotic explorations with artists-in-labs and currently she's preparing her new installation with robots and planning her upcoming projects.<http://www.alinamn.com>

PABLO VENTURA (ES/CH 1959) founded the Ventura Dance Company in London in 1986, for which he created numerous choreographies and videos in London, Madrid and Zurich. Other works have also included choreographies for contemporary operas, film, video and installations in collaboration with video artists, software designers, robot artists and electronic music composers. Ventura is the winner of various awards such as the *City of Zurich Dance Award* for his work in the field of dance and media in Zurich. Winner of the CYNETart prize for computer-created performances. He also received the dance prize of the year 2002 from the Canton of Zurich 'in recognition of his merits in dance-aesthetic research and innovative choreography'. This process culminated in 2006 with the commission of a work for a robot as the only actor in a dance piece; *Kubic's Cube*, developed during an artistic residence in Tesla-Berlin was presented at the Berlin's Transmediale in 2006. In 2007 Ventura was selected for the artists-in-labs program at the Artificial Intelligence Laboratory of the University of Zurich and was invited with *Kubic's Cube* to the International Symposium for Electronic Arts ISEA 2008 in Singapore (Lucid Fields). <http://www.ventura-dance.com>

DVD AND ANALYSES

CINEMA SOCIOLOGY Marille Hahne

Since 2003, I have been involved in the documentary and review aspects of the artists-in-labs program, and have involved Masters and ex-students from the Film Department at the Zurich University of the Arts in the production of films about this challenging and novel interaction between artists and scientists. As an engineer turned filmmaker myself, I am eager to be part of a team that appreciates the ability of the genre of 'documentary film' in order to collect notes about these interactions. This was accomplished by interviewing scientists and artists upon the completion of their residencies in the various science lab locations. We thought that these films would not only showcase the evidence of this interaction but, because of the cutting potential of film, could become a kind of moving image glue: a binder that linked different personalized reflections together.

The questions for the DVD films in the book were based on the advice of Priska Gisler, a sociologist from the Collegium Helveticum at the ETHZ, but this time we focused more on the discursive potentials and levels of interactions between the artists and the scientists themselves. By blending various answers to the questions from live interviews, details about the issues of know-how transfer, correlation, and potential collaboration, became apparent. The results can then be re-analyzed and used for evidence to support further research. So they are in fact, 'cultural biographies about cultural encounters'. These directions expand what we can imagine 'documentary' to be, even if they follow the rules of TV entertainment, because they study the realities of a residency encounter by engaging with and presenting the social interaction of people from very

different disciplines. As much of the framework for thinking about, making, and distributing documentaries has expanded over the last twenty years, the use of DVD docus to supplement theoretical and research oriented books like this one, can also help to expose where, and how the real and the fabricated have interacted with the artists-in-labs organizational structure and the resonances they accumulate.

The interviews are conducted in-situ and contain questions about the personal stories and motivations of the artist, the application procedure and their first month impressions as well as their encounters inside a science lab. They were also asked to describe their projects and the inspiriting moments, the scientists they met and the intensive nature of their exchanges. They were asked to comment on the style of the communication, their expectations, any critical or ethical reactions and the exchanges with other artists from other labs. In terms of know-how transfer, what did they learn? The artists' working processes and results were also described and filmed. In turn, the scientists were asked to comment on their motivations to take an artist into such an environment and the expectation they had imagined. They were asked to tell us how they had been involved, and if the experience resulted in any substantial collaboration with other scientists. What did they learn about art and would the artists leave any traces behind them when they left? Would they do it again and why?

By interrogating the encounter in such a way, the resultant films from this program can be used to gather an impression for many audiences (about scientific research in labs) from an 'outsiders' point of view. For here, the 'outsider' is the artist but also a 'narrator and catalyst' to introduce the audience into the complex world of science. Therefore, while genre might be closest to photography's use of historically conditioned codes of visual realism, it is one shared with a cultural narrator and the creative intention of the editor. This type of cinema sociology is designed for an adult community audience, and these films aim to expose the viewer to worlds beyond the orbits of personal experience.

CONTENTS: DVD ARTISTS-IN-LABS FILMS
DVD VIDEO PAL STANDARD PLAYABLE ON MAC, PC OR MULTICODE DVD PLAYERS

Labs and Artists

Life Sciences:
INSTITUTE OF INTEGRATIVE BIOLOGY (IBZ) | ETH ZURICH
_HINA STRÜVER & MÄTTI WÜTHRICH · 13 mins

CENTER FOR INTEGRATIVE GENOMICS (CIG) | UNIVERSITY OF LAUSANNE
_SYLVIA HOSTETTLER · 15 mins

WSL SWISS FEDERAL INSTITUTE FOR FOREST, SNOW AND LANDSCAPE RESEARCH, BELLINZONA
_CLAUDIA TOLUSSO · 12 mins

EAWAG: THE SWISS FEDERAL INSTITUTE OF AQUATIC SCIENCE AND TECHNOLOGY, DÜBENDORF
_PING QIU · 14 mins

Cognition & Physics:
THE BRAIN MIND INSTITUTE (BMI) | EPFL, LAUSANNE
_LUCA FORCUCCI · 12 mins

THE HUMAN COMPUTER INTERACTION LAB (HCI LAB) | INSTITUTE OF PSYCHOLOGY | UNIVERSITY OF BASEL
_MONIKA CODOUREY · 13.5 mins

PHYSICS DEPARTMENT AT THE UNIVERSITY OF GENEVA | CERN
_CHRISTIAN GONZENBACH · 13 mins

PAUL SCHERRER INSTITUTE (PSI), VILLIGEN
_ROMAN KELLER · 10.5 mins

Computing & Engineering:
CSEM SWISS CENTER FOR ELECTRONICS AND MICROTECHNOLOGY, ALPNACH
_PE LANG · 9.5 mins

THE NATIVE SYSTEMS GROUP | COMPUTER SYSTEMS INSTITUTE | ETH ZURICH
_CHANDRASEKHAR RAMAKRISHNAN · 10 mins

ISTITUTO DALLE MOLLE DI STUDI SULL'INTELLIGENZA ARTIFICIALE (IDSIA), MANNO-LUGANO
_ALINA MNATSAKANIAN · 11 mins

ARTIFICIAL INTELLIGENCE LABORATORY | UNIVERSITY OF ZURICH
_PABLO VENTURA · 11 mins

Credits
DIRECTOR MARILLE HAHNE
PRODUCER ARTISTS-IN-LABS, ZHDK
PRODUCTION MASTER OF ARTS IN FILM, ZHDK
COORDINATION KARIN RIZZI
INTERVIEWERS MARILLE HAHNE, IRÈNE HEDIGER, JILL SCOTT
CAMERA ANDREAS BIRKLE, THOMAS ISLER
EDITORS ANNETTE BRÜTSCH, MARILLE HAHNE, THOMAS ISLER, ROMANA LANFRAN-CONI, ROLF LANG, CLAUDIA LORENZ, ALAIN RICKLI, FRANZISKA SCHLIENGER
SUBTITLES MARILLE HAHNE
ONLINE EDIT AND AUTHORING NICO LYPITKAS, RONNIE WAHLI
SOUNDMIX GREGG SKERMAN, FILMSTUDIES, ZHDK
DVD PRINT DIGICON

ALL FILMS ARE IN ENGLISH OR HAVE ENGLISH SUBTITLES